Oil Spill Studies

To my mother and my father

Oil Spill Studies

Healing the Ocean,
Biomarking and the Law

Edited by

Frédéric Muttin

First published 2018 in Great Britain and the United States by ISTE Press Ltd and Elsevier Ltd

ISTE Press Ltd
27-37 St George's Road
London SW19 4EU
UK

www.iste.co.uk

Elsevier Ltd
The Boulevard, Langford Lane
Kidlington, Oxford, OX5 1GB
UK

www.elsevier.com

British Library Cataloguing-in-Publication Data
A CIP record for this book is available from the British Library
Library of Congress Cataloging in Publication Data
A catalog record for this book is available from the Library of Congress
ISBN 978-1-78548-310-3

Printed and bound in the UK and US

Contents

Preface

The oceans of tomorrow will suffer from many human pressures and uses. This book presents interdisciplinary fields and actions concerning the response to chronic and accidental marine pollution, including oil spills. The objectives appear so immense that this preface aims to ascertain the orientation chosen for the scientific and technical works presented.

Intense media coverage of marine pollution can be very sporadic. The previous decades saw the vast amounts of pollution sources identified, among other things, as plastic soup, lost containers, heavy metals and hydrocarbon leakages. By contrast, the many different research activities take many years to achieve the ongoing knowledge progress. For example, we need to address various maritime environments and many kinds of pollutants and polluters encountered in different spatial and temporal situations. The first scissors effect between the step-by-step progression of research and the large public involvement led us to present various situations and research methodologies in order to show the complexity of elaborating several scientifically based responses. For example, Chapters 1 and 2 introduce seven different types of maritime zones and land fields, and investigate a technological response based on floating barriers that can be used in an emergency situation.

In the last decade, the public has met with another kind of scissors effect. First, this effect results from the increase in shipping activities and transportation as well as deep-sea mining exploration. Second, the excessive use of sea biodiversity for food capture and aquaculture as well as for new welfare products such as tourism has created a rift, leading to human ill-being as a result of contradictory demands on the marine and ocean environment. To this end, Chapters 3 and 4 present research activities on ecotoxicology that focus on different types of chronic water pollution. This is explained by showing a yachting marina and mid-depth fish contamination by hydrocarbon dispersants that are used during a leakage of deep-sea oil wells.

In recent years, new kinds of crises have resulted from different pollutant origins. The length of a polluted coast can reach several thousands of kilometers, while the pollution source in the sea can be deeper than 1,000 m. New Arctic shipping routes and recent oil exploration can also pose new challenges. Chapter 5 addresses the extreme environments, particularly in Arctic and cold water regions or in the deep sea with high water pressure.

Finally, Chapter 6 deals with the dimension of the Law and the Rights concerning the sea environment. Each marine and coastal activity is linked to a juridical constraint. Therefore, we need to open up a discussion on this problem in order to obtain a valuable vision of the marine pollution subject.

Frédéric MUTTIN

June 2018

Acknowledgments

We would like to acknowledge the following people and institutions:

– The *Secrétariat général de la mer* (SGMer), Marie-Sophie Dufau-Richet, Paris.

– The European Commission DG-ECHO, Humanitarian Aid and Civil Protection, and the Civil Protection Financial Instrument, Brussels.

– The INTERREG Atlantic Area Transnational Program ERDF (the European Regional Development Fund) and the *Centro Tecnológico del Mar Fundación* (CETMAR), Vigo.

– The EIGSI Engineering School La Rochelle and Casablanca, Joël Jacquet, Director of Research.

– The University of La Rochelle and the Research Unit "Littoral, Environment and Societies" (LIENSs).

– The Cedre experts center, Brest.

Our acknowledgments also go to all those who contributed to this book, in particular Thomas, Frédérique and David.

Author Biographies and Organization Presentations

The following are short descriptions of the authors' organizations. The biography of each author is also given.

ALLEGANS institute

ALLEGANS institute is a network of specialists in legal issues focused on managing the different organizations concerned. This body works on the accidental effects and chronic contamination of coastal areas. These issues are notoriously difficult to be implemented by public laws, decision-makers and management tools. One of their main activities is to link biomonitoring with pollution in order to manage the causes and effects of such a contamination. Indeed, assessing the consequences of accidental and chronic contamination in coastal areas is a difficult task. ALLEGANS institute aims to compensate for and anticipate future pollution.

Yann Rabuteau, Founder and Manager of ALLEGANS institute, contributed to *Chapter 6* of this book.

Yann RABUTEAU is a lawyer specialized in maritime and coastal topics, corporate relations, research and innovation.

He is a graduate in Public Law from the *Université de Bretagne Occidentale* (UBO) and specialized in the Law of the Sea, maritime activities and marine environment. He is also an associated researcher at the AMURE Institute (*Aménagement des usages des ressources et des espaces marins et littoraux UMR CNRS 6308*) where he teaches environmental legislation on pollution. He is a parliamentary attaché of Brittany, Brest district.

CEDRE association

The CEDRE is the Centre for Documentation, Research and Experimentation on Accidental Water Pollution, located in Brest, France. It is a nonprofit organization founded on January 25th, 1979, as part of the measures taken in the aftermath of the Amoco Cadiz oil spill. Its missions are to provide advice and expertise to the authorities responsible for the spill. It is a national competent authority for both marine and inland surface waters. It continues to gain expertise and develop tools in order to fulfill its various duties. Its advice and skills can also be aimed at foreign authorities or private companies.

Stéphane Le Floch, Research Department Manager of the CEDRE association, co-wrote *Chapters 4 and 5* of this book.

Stéphane LE FLOCH is a chemist by training and has been working at CEDRE since 1995, but acted as a supervisor on behalf of the oil company ELF Petroleum Norge in Norway between 1998–2000. In 2014, he was appointed Manager of the Research Department at CEDRE. Since January 2008, he has been the French representative of the GESAMP working group at the International Maritime Organisation (IMO). This group is in charge of the Evaluation of the Hazards of Harmful Substances Carried by Ships (GESAMP/EHS). Since 2017, he has also been involved with a working group in charge of hazardous and noxious substances (HNS) topics for the European Maritime Safety Agency (EMSA).

EIGSI Engineering School

Founded at the beginning of the 20th Century in Paris, EIGSI Engineering School now has two campuses, one in the La Rochelle city on the French Atlantic coast and the other in Casablanca, Morocco. Future engineers graduate from EIGSI at the end of a five-year program. Three hundred Post-Bac students in Casablanca and a thousand students and apprentices in La Rochelle receive their degree in General Engineering Sciences and Humanities. EIGSI was awarded both the EESPIG label as a public–private institution and the EUR-ACE European label. It is an institution linked to the district council *Conseil-Départemental de la Charente-Maritime*. The three research topics of EIGSI are Urban Mobility, Renewable Energies and Coastal Protection. Activities in marine and coastal environments concern accidental and chronic marine pollution. Numerical models were studied and full-scale experiments on fluid–structure interaction were carried out. The structures were made up of flexible fabric materials, cables or membranes. The fluid flows of marine and coastal environments were studied in semi-open or open areas, natural sites, estuaries, rivers, ports and terminals.

Rose Campbell, an oceanographer, co-wrote *Chapters 1 and 2* of this book.

Frédéric Muttin, a researcher and Professor of Applied Mathematics, co-wrote *Chapters 1, 2 and 6* of this book.

Rose CAMPBELL is a researcher in the field of marine sciences. Her research interests include numerical modeling, optimization, oceanography and software development. She holds a PhD and Masters in Oceanography from Aix-Marseille University, France, and a bachelor's degree in Science/Mathematics from the University of Ottawa, Canada. She currently teaches at EIGSI Engineering School in La Rochelle, France. During her PhD studies, she contributed to a French national research project on the study of mesoscale currents (LATEX, ANR funding). More recently, she has contributed to several European research projects related to marine pollution (ISDAMP, Improving Shorelines Protection Against Marine Pollution, and ARCOPOL-Platform) and marine renewable energy (EnergyMare and Turnkey). The philosophy of these projects is that local stakeholders should have access to the best numerical tools to make decisions about the use of anti-pollution equipment and the installation of MRE technology.

Frédéric MUTTIN has worked as a Professor of Mathematics for 25 years. He obtained his PhD in Applied Mathematics from the University of Nice-Sophia-Antipolis, France in 1989 and the Habilitation HDR from the University of Limoges, France in 2010. His research activities include study on marine environment, coastal protection and fluid–structure interaction at EIGSI Engineering School in La Rochelle, France. Recently, he obtained funding for the ISDAMP (coordination) and ARCOPOL-Platform “Maritime Pollution Preparedness and Response in Atlantic Regions” projects from the European DG-ECHO and INTERREG Atlantic Area. His interdisciplinary research includes applied mathematics and environmental sciences on the preparedness for oil spills which threaten coastal areas and harbors. He now manages a group of students in EIGSICA Casablanca, Morocco, focused on innovation and entrepreneurship toward a geometrical level-set method applied to a harbor environment and different optimal processes.

LIENSs laboratory

LIENSs (Littoral Environment and Societies UMR CNRS 7266) belongs to both research authorities, CNRS and La Rochelle University. It is composed of joint research units and researchers having different scientific domains. The sciences interdisciplinary components are the Environment (biology, ecology, ecophysiology, ecotoxicology, Earth sciences, geophysics), Humanities and Social Sciences (geography and history), and Engineering (chemistry and biotechnology). Principal research topics address the interdisciplinarity and sustainable development of the littoral zone.

Marine Breitwieser, a PhD ecotoxicologist researcher, co-wrote *Chapter 3* of this book.

Thomas Milinkovitch, a marine biologist and ecotoxicologist, co-wrote *Chapter 4* of this book.

Hélène Thomas-Guyon, Manager of Marine Animals Responses to Environmental Variability Team, co-wrote *Chapters 3 and 4* of this book.

Marine BREITWIESER is a PhD researcher in Ecotoxicology at LIENSs laboratory in the Marine Animals Responses to Environmental Variability (AMARE) team. Together with La Rochelle Harbor, and in particular Angélique Fontanaud, she conducted a first biomonitoring study on harbor environmental quality during her end-of-course internship. She has improved skills in many scientific fields in semi-closed and open areas (harbor and coastal environments). During her PhD, she collaborated with geophysics researchers to explore and explain chemical effects on biodiversity, and more particularly, the environmental quality with guidelines and normalized tools (chemistry, biochemistry).

Thomas MILINKOVITCH obtained his PhD in Biological Oceanography and Marine Environments from the University of La Rochelle, France. His doctoral works focused on the toxicity of oil dispersants during oil spill disasters. More specifically, he studied the impacts of dispersed hydrocarbons on the metabolism, disease and oxidative stress of the golden grey mullet. During his postdoctoral studies, he investigated the impact caused by hydrocarbons and some heavy metals on two species of bivalve and pectinidae (*Chlamys varia* and *Chlamys islandica*). More recently, he conducted research on the impact of hypoxia and dispersed oil at the *Consiglio Nazionale delle Ricerce* (CNR) in Italy, in order to show the effect they had on sea bass. These two stress factors influenced the metabolism and behavioral patterns of this species.

Hélène THOMAS-GUYON is a researcher and lecturer (HDR *Maître de Conférences*) at La Rochelle University, France. She heads the research group AMARE of LIENSs laboratory (Littoral Environment and Societies) unit no. UMR-7266 of the French CNRS. Her research focuses on the physiological and immunological responses of marine organisms to external pressures (e.g. pollution, temperature) or internal factors (e.g. size). Her research interests are eco-physiology, immunity and mechanisms of adaptation. She has participated in European collaborative research through projects such as INTERREG IVB Atlantic Area ANCORIM (2009–2012) and NanoReTox FP7-NMP (2009–2012). She has also led National CNRS projects such as PEPS MAPS RISPECT (2014). She is currently heading the La Rochelle Harbor project (2015–2018) as well as two international projects, PECTIMPACT and EPOLSCALL, with the Royal Norwegian Embassy and the Ministry of Foreign Affairs (2014–2019).

LABOCEA

LABOCEA is a public analysis laboratory located at five sites in Brittany, France. Its fields of expertise range from the analysis of major pollutants (e.g. hydrocarbons, pesticides, microplastics, endocrine disruptors, etc.) to the microbial characterization of contaminated microbiota and aquatic ecosystem survey. This laboratory fulfills a public service mission designed to meet the needs of the state, local authorities, professionals and individuals. It is the most important structure in France in terms of skills, technical facilities and human resources.

Florian Lelchat, a marine microbiologist, co-wrote *Chapter 5* of this book.

Florian LELCHAT is a researcher with a PhD in marine microbiology. His research interest involves the study of interactions of marine bacteriophages with their environment from an abiotic and a biotic point of view. He has also been involved in bioprospecting for two bacteriophages and bacteria, especially in the Antarctic on psychrophilic microorganisms. He also has a rich background in metabolomics, proteomics and biotechnologies. He is currently posted at LABOCEA, Plouzané, France.

LMGE laboratory

LMGE (*Laboratoire Microorganismes, Génome et Environnement* UMR CNRS 6023) conducts research on microorganisms (prokaryotes, eukaryotes and viruses), from molecular and cellular aspects to the roles of these organisms in ecosystems. The *Interactions dans les réseaux trophiques aquatiques* (RTA) team at LMGE focuses on diversity at different levels (infraspecific, specific and functional) of microbial food webs in the aquatic environment. Using both ecosystem and mesocosm approaches, the RTA team studies trophic links between microorganisms to highlight the diversity of energy pathways and the impact of the alteration of this diversity on the regulation capacity of the systems.

Matthieu Dussauze, a marine scientist, co-wrote *Chapter 5* of this book.

Matthieu DUSSAUZE is a postdoctoral researcher at Akvaplan-niva AS, Tromsø, Norway, who works on multi/interdisciplinary research projects and has investigated areas such as marine biology, ecotoxicology and physiology. He has specialized in the biological impact of oil spills on several ecosystems including temperate, Arctic and deep-sea ecosystems. He has published 18 peer-reviewed papers, mainly on oil spill impacts on fish species at several levels of organizations (from mitochondria to population). He holds an individual fellowship from the European Commission, Marie Skłodowska-Curie Actions, and continues his research between Norway at Akvaplan-niva AS, and France at the University of Western Brittany.

La Rochelle Harbor

La Rochelle Harbor was created in 1972. It is one of the largest harbors and the biggest on the Atlantic coast with 5047 places to keep boats laid out across 70 hectares. This harbor has been known for its environmental management system since 1985 with the Blue Flag Label and the ISO norm 14001 since 2006, and for its developing actions with the researchers of LIENSs institute. Bertrand Moquet is the harbor manager.

Angélique Fontanaud, an Environmental Department Manager, co-wrote *Chapter 3* of this book.

Angélique FONTANAUD is the Environmental Department Manager of La Rochelle Harbor. In this harbor, since 2006, she has developed an environmental management system according to the ISO norm 14001. Moreover, she has developed biomonitoring actions with the researchers of LIENSs institute, in particular Hélène Thomas-Guyon and Marine Breitwieser.

1

Oil Spill Containment in Semi-open Areas: Experiments in French Atlantic and Alpine Waters

In this chapter, experiments conducted on the mechanical containment of oil pollutants in semi-open areas are presented. Located on the French Atlantic coast of the Charente-Maritime district and the Maurienne valley in the Alps, the experiments demonstrate original contingency plans with different geometries and methods to control oil release. These coastal and inland sites are subject to specific environmental constraints and significant damage if oil containment fails.

1.1. Introduction

Many strategies are available to combat oil spills using floating barriers [FIN 16]. Located in Rochefort, La Rochelle and Hermillon, France, the following three studies used oil spill booms in different hydrographic and marine environments. Table 1.1 presents the details of the strategies used in the experiments. Different shapes of boom deployment are considered as follows: a line with moored end-points, a closed line such as a triangle and a chevron.

Chapter written by Frédéric MUTTIN and Rose CAMPBELL.

The head of the chevron can be positioned upstream to deviate oil on each side or downstream to accumulate oil.

The position of anchor points can be time dependent due to forcings such as currents and wind or due to an intervention strategy. A static intervention means that a boom part is generally immobile. On the contrary, a dynamic intervention involves displacing a boom end-point generally using a ship.

Experiment location	**Boom geometry**	**Intervention strategy**	**Environment**	**Working objectives**
Rochefort Harbor	Line	Static at quay Dynamic	Estuarine harbor	Sweeping Concentration
Chef-de-Baie Harbor	Triangle	Static	Oceanic harbor	Containment Absorption
Maurienne Lake	Chevron Double-head	Static	Freshwater lake	Exclusion Diversion

Table 1.1. *Summary of three boom installations in the semi-open areas of France*

The objectives of the floating barriers used are to collect, deviate and displace floating pollutants, which vary between the sites. For each site, modeling has been implemented to carry out re-analysis of real scenarios.

1.2. Rochefort Harbor

Different responses to small-scale pollution can be adapted to the requirements of harbors [KRE 07]. This experimental scenario is based on a real accident where a shipwreck was dismantled inside the basin of the river port of Rochefort, leading to a small discharge of oil. At that time, responders used oil-absorbent materials which were directly disposed on the surface of the basin. No efforts were made to contain or concentrate the slick. Responders considered closing the gates between the basin and the river in order to confine the pollution. The drawback of this action is that the commercial shipping activity of the harbor would be interrupted until the pollution was resorbed.

The experimental method used a 50 m curtain boom as a sweeping device on the polluted water surface. A small boat was used to sweep one end-point of the boom in a semicircle, in order to concentrate the hypothetical pollutant near a mooring bollard on the quay. The other end-point of the boom was moored on a quay ladder and equipped with a dynamometer and tension recorder to prevent boom and mooring failure. This scenario considers less than 1 ton of oil pollutants following an accidental release during bunkering from a truck or the breakdown of an old ship.

The towing velocity of the dynamic boom end-point remained in the order of one knot, taking 3 to 4 minutes to complete one arc (Figure 1.1). This method concentrates the pollutant along the quay, where an absorbent material or a skimmer can be used to absorb the oil.

Figure 1.1. *Rochefort basin and sweeping boom, December 5th, 2014*

A total of 13 people took part in this experiment: local responders, researchers, students and port authorities. The participant organizations were the Engineering School of La Rochelle, the port of Rochefort and Tonnay-Charente, the Charente-Maritime council and the port of Falmouth.

The boom position was recorded using complementary instruments. A GPS device recorded the position of the end-point located near the boat. Two video cameras filmed the sweeping process from two boundary points on the quay. An overhead view of the boom was obtained using image transformation and georectification. Finally, a flag was mounted on the boom, and three observers tracked the flag throughout the experiment and noted its azimuth relative to the quay using a large protractor for every 20 seconds. The position of the flag was deduced using triangulation, in order to confirm and calibrate the georectified image results. The positions of all of the scientific observers relative to the boat and the boom are shown in Figure 1.2.

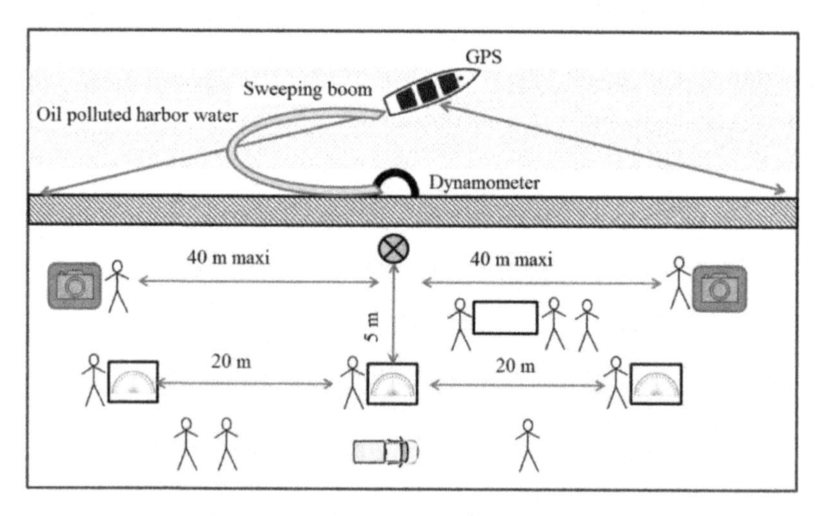

Figure 1.2. *Semi-dynamic sweeping boom for oiled water in the harbor*

The time taken to prepare and deploy the boom and the boat was 30 minutes. A total of 14 sweeps were carried out, seven from left to right and seven from right to left. In order to give clear and precise navigational instructions to the boat's pilot, the theoretical semicircular trajectory was replaced by a triangular sweeping of the basin. The triangular movement requires only one change of direction of the small towing boat at the middle of a run (consequently, only two azimuths are communicated to the pilot).

The GPS coordinate data give a valuable boat position in real time. A forward numerical derivative of these boat coordinates delivers a first-order approximation of the boat velocity. Three velocity singularities arise during a run, at the starting and ending points where the initial and final velocities are null, and at the turning mid-point where the velocity direction is discontinuous. The duration of run number 7 was 4 minutes with a mean boat speed of 0.55 m/s (Figure 1.3).

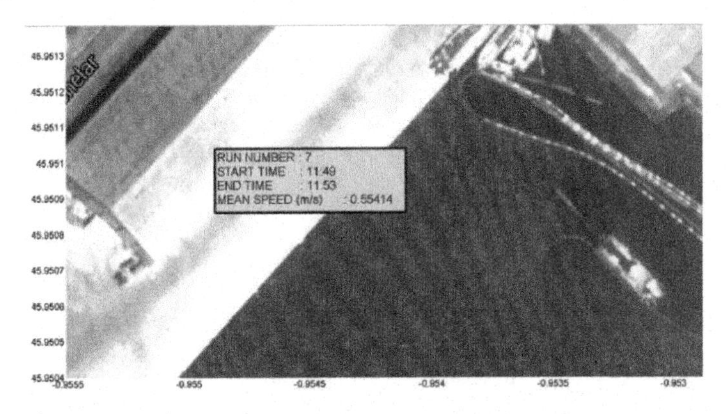

Figure 1.3. *Boat velocity (red vectors) as deduced by the first-order derivative of GPS coordinates. Background: Rochefort Harbor basin as seen in Google Maps. For a color version of this figure, see www.iste.co.uk/muttin/oil.zip*

The boom has a straight geometry only at the starting and ending points of the run. Two numerical models are used to approach the boom motion during a run. We use both 2D cable and 3D membrane finite-element models [MUT 17]. The boom behaviors at the 7th time step of the run by using these two models are shown in Figure 1.4. This figure is a screenshot of the BARRIER software.

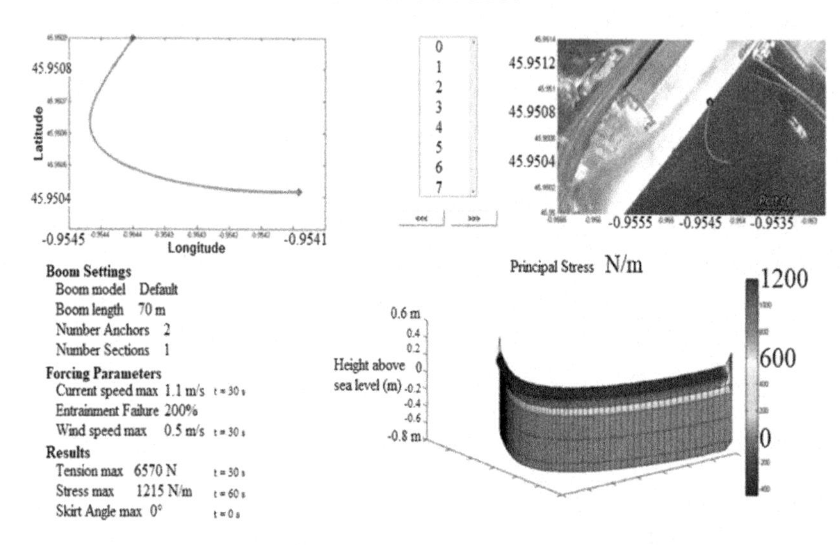

Figure 1.4. *Mechanical modeling of semi-dynamic sweeping with 3D finite-element giving boom tension, skirt angle, fabric stress and oil leakage criteria. For a color version of this figure, see www.iste.co.uk/muttin/oil.zip*

In case of pollution, the Rochefort Harbor must respect national legislation by allowing the fire brigade SDIS to carry out the response action using their resources, equipment and personnel. Nevertheless, the harbor master and the harbor owner, the Charente-Maritime district council, have been favorable to research work that involved testing new ideas. The current experiment has allowed a fruitful collaboration between scientists and local authorities, the office of maritime affairs of the district council. The Mayor of Rochefort and Vice-President of the district council have made efforts to endorse environmental protection and marine activity development.

The following section will extend the research work to another port of the Charente-Maritime district under another local chain of command.

1.3. Chef-de-Baie Harbor

A place of refuge for small fishing vessels (<12 m) is located inside the fishing harbor of Chef-de-Baie on the French Atlantic coast. Chef-de-Baie Harbor is home to approximately 50 small boats (<12 m) and fewer than 10 larger boats (>12 m). Each year, accidents in the Charente-Maritime district occur when boats are towed to a sheltered place. This harbor is used because it is semi-open and accessible at all tides from the ocean.

The damaged boat is placed near a quay with an oil spill boom stretched around its waterline to prevent pollution from leaking inside the harbor basin. Several buoyancy bags can be used to sustain the boat. Divers, small motor boats and a crane are typically used to lift the damaged boat onto the quay.

The experimental pollution response protocol involved the installation of a triangular boom around the hull of a hypothetical ship (Figure 1.5). Each side of the triangle measured 10 m, using two boom curtain sections and one absorbent boom section. The containment sections limited oil dispersion from the leaky hull, even if the hull sank or was elevated by the crane. The absorbent section was placed downstream of the dominant oil drift movement created by currents, waves or wind. To handle the triangular boom around the wreck, three mooring lines of 5, 10 and 18 m were placed at the triangle corners and moored on bollards. To avoid using boats and divers, all manipulations were carried out by hand, using only two responders.

The triangular boom was deployed from a floating dock in the harbor, moored once on the dock and moored twice on the quay where cranes were typically positioned (Figure 1.6). The exercise that took place on March 16th, 2015 was carried out by the Engineering School of La Rochelle for the harbor owner, the Charente-Maritime district council.

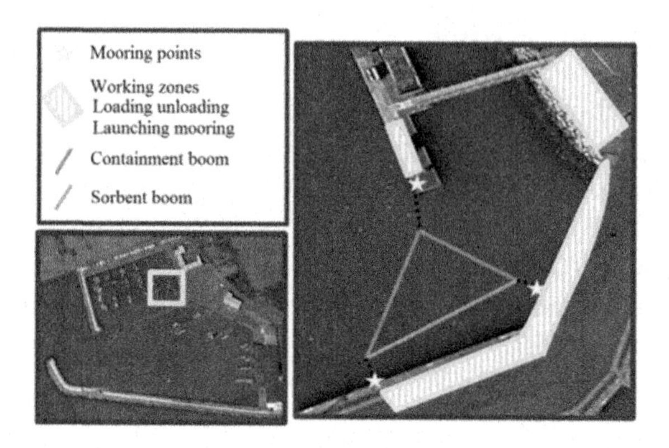

Figure 1.5. *Simulating containment of leaky ship experimental protocol. Left: the location of the exercise area inside the harbor basin. Right: the working zones where the boom was deployed and the mooring lines were managed to appear on the dock and quay. For a color version of this figure, see www.iste.co.uk/muttin/oil.zip*

Figure 1.6. *Boom deployment and mooring exercise. Chef-de-Baie fishing harbor, March 16th, 2015. Left: fish freight handling quay. Right: fueling station*

Let us consider the physical-chemical properties of a typical pollutant in this harbor (Table 1.2). The diesel marine fuel for the fishing boat is called Gazole-Peche with a flash point of 60–120°C. The surface tension properties of Fuel no. 2 are taken into account.

Property	**Temperature (°C)**	**Value**
Oil reference		Gazole-Peche
Kinematic viscosity (cSt or $mm^2.s^{-1}$)	40	3 ($3\ 10^{-6}\ m^2.s^{-1}$)
Dynamic viscosity (Poiseuille, i.e. Pa.s)	40	0.252 (2.52 mPa.s)
Density	15	0.84
Oil–saltwater surface tension ($N.m^{-1}$)	15	0.0136
Oil surface tension ($N.m^{-1}$)	15	0.0274
Oil–fresh water surface tension ($N.m^{-1}$)	15	0.0147

Table 1.2. *Marine fuel oil properties*

The boom was deployed for 5 hours, representing a tidal amplitude of 250 cm. Mooring lines were regularly adjusted to account for the varying water levels, and current and wave actions were negligible throughout. Overall, the boom retained its triangular position, and the mooring lines were unstressed (Figure 1.7).

An important aspect of preparing an experiment or exercise is to anticipate meteorological and oceanic conditions. For this exercise, oceanic waves that can be reflected inside the harbor basin when coming from the North-West were of particular concern. Such wave patterns can hinder the pollution response in the exercise zone. To this end, a wave modeling system was implemented using the Wavewatch 3 and Swan models. These models were implemented on a multigrid downscaling system using three embedded mesh grids in the La Rochelle region [ASC 15]. A drawback of these models is the quality of the grids. Pollution can occur in recently anthropized coastal zones for which there are no latest shoreline data.

Such modeling gives the dominant swell direction outside the harbor, but there is no refraction or other processes within the basin.

Figure 1.7. *Triangular boom plan at the port of refuge in Chef-de-Baie fishing port, La Rochelle, March 16th, 2015*

The wave forcing factor is modeled in the coastal region of the Chef-de-Baie Harbor [FRA 14]. The wind and the current actions on the triangular boom plan are considered in a 2D cable model.

The model is a first-order differential equation that gives the tangential boom tension T in terms of the curvilinear coordinate s along the elastic domain. The external action on the boom is a normal pressure p that belongs to the normal vector n(s) at the curvilinear position in s:

$$\frac{d}{ds}T(s) = p.n(s) \quad [1.1]$$

The equation is nonlinear due to the dependence on the tension vector T with the unknown equilibrium position of the boom between external and internal forces.

In the modeled example, the current is parallel to the rocky shore (Figure 1.8). The modeling results and forcing vector map are superimposed on the aerial photo of the basin. This part of the contingency plan helps the responders to adapt the mooring to the wreck position.

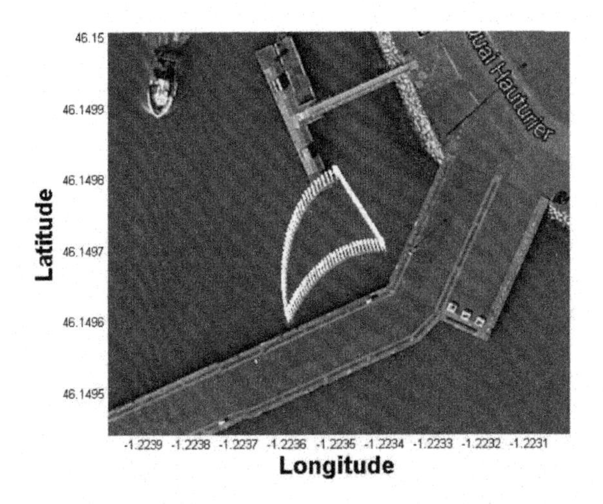

Figure 1.8. *Modeling the response of triangular boom plan to an along-shore current in a semi-open harbor*

The following section continues the discussion from oceanic to mountain lake conditions.

1.4. Maurienne Hermillon Lake

In autumn 2014, an accident occurred in the French Alps. Heavy oil was released into a hydroelectric power plant and flowed through a series of cascading artificial lakes. A containment boom was installed in Hermillon Lake to control the floating part of the pollutant [DOL 14].

The presence of oil in this hydraulic system is exceptional, and, as such, no specific contingency plan was available to deal with hydrocarbon pollution. As a result, the installed oil containment plan was based on the responder's experience in different fields. The national authority defined the objective as to contain the oil pollutants to preserve the downstream Isère River.

The lake dimensions were 530 m long and 350 m wide. A decision was made to install an oil spill boom of 245 m long. The boom plan consisted of a barrier and a curtain with inflated float. During installation, the water level in the lake was 0.5 m (Figure 1.9).

Figure 1.9. *Boom deployment in Hermillon artificial lake before the increase in water levels from 0.5 to 11.5 m in 24 hours*

Two responders were involved in this intervention, from the national authority, fire brigade and oil spill stockpile. The response material was carried by a truck to the site and with a pneumatic boat on the lake.

The hydraulic plant was restarted after the oil spill boom became operational. After 24 hours, the water level in the lake reached 11.5 m due to the accumulation of upstream potentially polluted waters. When lakes are close to capacity, there is a risk of releasing oil into the weir, and operators and materials can fall into the spillway (Figure 1.10). Under this condition, any small increase in the water level could entrain the floating pollution and working boat into the spillway.

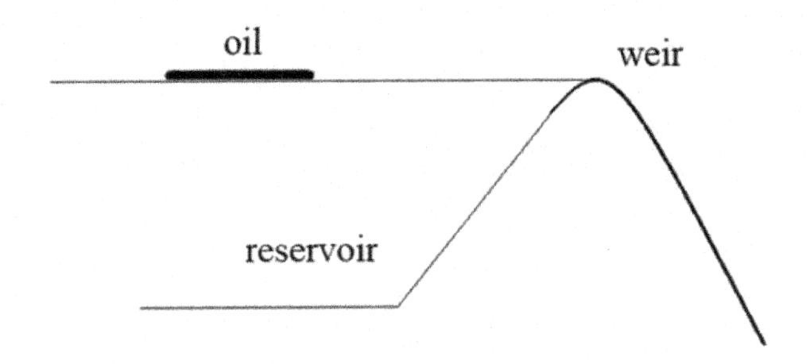

Figure 1.10. *Oil pollutants in the reservoir drifting to the weir*

Figure 1.11 shows the double chevron geometry of the installed boom. Two anchors weighing 35 kg each were moored at the middle points of the chevron. The mooring lines of the anchors consisted of a 20 m polypropylene line of diameter 22 mm and a 10 m chain of diameter 12–10 mm. The total weight of each mooring line was 50–60 kg, including the anchor weight. The bottom of this artificial lake was a flat concrete floor covered with bitumen. The anchors were installed in a mud flat layer with a thickness of 0.5 m.

The end-points of the chevron were attached with a stretched cable to the lake bank. A tide compensator device was installed at both end-points to adapt the boom to changing water levels.

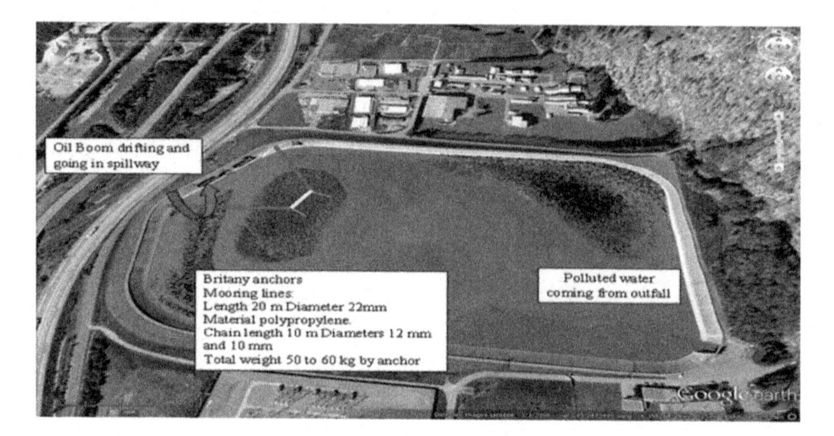

Figure 1.11. *Oil spill boom plan in the double chevron with two mooring lines and anchors, including the pollution source from the outfall and possible drifting damage of the material into the spillway*

The initial water flux in the lake entrance was 40–60 $m^3.s^{-1}$, and the boom plan was effective at containing floating oil. Later on, a decision was made to increase the flux to 70–90 $m^3.s^{-1}$. As a result of the rapid flux increase, a wave traveled on the surface from the outfall entrance to the boom plan. Under these hydrodynamic conditions, anchors drifted on the lake bed and the boom plan was displaced over the weir and fell into the spillway.

To analyze the failure of this boom plan, a modeling approach was used in a hindcast. Hermillon Lake is known as the Longefan reservoir by the national electric company. A flushing of this reservoir was carried out to study the sediment control inside the whole hydraulic system of the Arc en Maurienne River. For a water flow of 70 $m^3.s^{-1}$, the experimental and numerical results give a water current velocity ranging from 1 to 1.3 $m.s^{-1}$ [CAM 11]. The present study used the velocity vector field that occurred during the sediment flushing into the Longefan reservoir on September 25th and 26th, 2008. The maximal water current velocity considered was 1 $m.s^{-1}$. The wind velocity was considered to be null. The simulated results are shown in Figure 1.12.

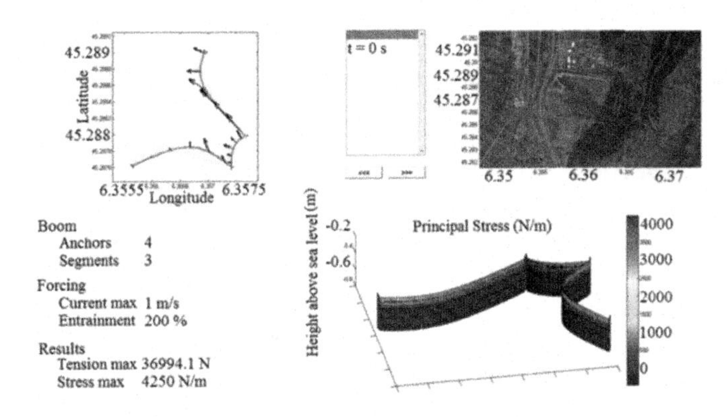

Figure 1.12. *BARRIER oil spill boom computations with 2D and 3D finite elements indicating stressed geometries and fabric tension. Upper right: equilibrium geometry of the boom chevron on a satellite image of the lake. Upper left: 2D initial and equilibrium geometries together with the current velocity vector field. Lower right: 3D boom geometry with the color map indicating the fabric stress intensity. Lower left: summary of computation results. For a color version of this figure, see www.iste.co.uk/muttin/oil.zip*

The oil entrainment was twice the failure threshold, suggesting that floating oil can flow under the boom for the surface current considered. The boom tension reached 36,994 N. The tension amplitudes on the mooring lines over the reservoir floor were very high and appeared larger than the anchoring resistance on the mud flat. We note that the maximal flux considered in the reservoir flushing study was 70 $m^3.s^{-1}$. During the pollution response, the flux reached 90 $m^3.s^{-1}$, suggesting the possibility of sediment erosion. The anchors were moored on the top of the mud flat. The sediment might have been detached by the water flow; as a result, the anchors would have been ripped. The water level into the reservoir might be lower during the flushing operation than during the pollution response. Using a 3D hydrodynamic model of the lake could quantify the water velocity at various depths in the reservoir. The jet direction of the flow at the reservoir entrance influences its sedimentation. During a future pollution event, the jet direction must be precisely monitored for the safety of the intervention and accurately measured to ensure the prediction of quality boom models.

1.5. Discussion

A narrow water exhaust is a common property of river port basins and artificial lakes. Closing this exhaust by gates, generally used for controlling the water level, can shut these semi-open areas for the flow of floating pollution. On the contrary, the basin of a fishing harbor such as Chef-de-Baie remains open and physically connected to the ocean all the time. In such a case, during the entire pollution event, a closed loop around a floating pollution is recommended to contain it.

Table 1.3 presents the main factors that summarize the experiments presented in this chapter. The first two factors, the number of people involved and the boom length, show the physical and financial efforts necessary for each experiment. The last two factors, the maximal flow velocity and the failure of the boom plan, show the difficulties encountered.

Experiment location	Number of people involved	Length of boom (m)	Maximal flow velocity ($m.s^{-1}$)	Failure of boom plan
Rochefort Harbor	13	50	1.5	NO
Chef-de-Baie Harbor	2	30	0	NO
Maurienne Hermillon Lake	2	245	1	YES

Table 1.3. *Main factors of boom installations in semi-open areas*

The number of people involved in an experiment is an interoperable piece of data that allows valuable exchanges between stakeholders and the chain of command. The maximal flow velocity during the experiment is a principal risk of boom plan failure. Generally, this risk increases when historical data and modeling results are unavailable.

In the three regions, fishing is an important socio-economic activity, which reinforces the need to conduct experiments on the pollution response. The protection of the biodiversity and fish resources are addressed in the Charente, Arc en Maurienne and Isère rivers. The commercial activity and dock handling chain are the beneficiaries of oil pollution recovery in the fishing harbor of Chef-de-Baie.

1.6. Conclusion

The experiments described in this chapter consider pollution in close proximity to the location of the oil release. When the pollutant extends to a wide zone, the methodology may be preceded by an environmental sensitivity index analysis. The most sensitive zones can be prioritized for the deployment of response equipment [IBA 17].

This chapter demonstrates the positive impact of using mechanical modeling of the floating barrier. Good quality data on physical forcing provides valuable model results on oil spill booms. These results can be included in a contingency plan to guide responders for a confident deployment of resources.

The local and district levels in the chain of command are considered in the inland and shoreline interventions presented in this chapter. Establishing reactive responses to moderate-sized pollution events can be done by integrating research activity into the preparedness activities of local and district authorities.

1.7. Bibliography

[ASC 15] ASCIONE-KENOV I., MUTTIN F., CAMPBELL R. *et al.*, "Water fluxes and renewal rates at Pertuis d'Antioche/Marennes-Oléron Bay, France", *Estuarine, Coastal and Shelf Science*, vol. 167(A), pp. 32–44, 2015.

[CAM 11] CAMENEN B., PAQUIER A., BOUARAB A. *et al.*, "2DH modelling of a reservoir flushing compared with LSPIV measurements", *34th IAHR World Congress June 2011*, Brisbane, Australia, 2011.

[DOL 14] DOLLHOPF R.H., FITZPATRICK F.A., KIMBLE J.W. *et al.*, "Response to heavy, non-floating oil spilled in a Great Lakes River Environment: a multiple lines of evidence approach for submerged oil assessment and recovery", *International Oil Spill Conference Proceedings May 2014*, vol. 2014, no. 1, pp. 434–448, 2014.

[FIN 16] FINGAS M., *Oil Spill Science and Technology*, 2nd ed., Gulf Professional Publishing, Cambridge, MA, 2016.

[FRA 14] FRANZ G., PINTO L., ASCIONE I. *et al.*, Wave modelling system implementation, ISDAMP project report, 2014.

[IBA 17] IBARRA-MOJICA D., ROMERO Á., BARAJAS-FERREIRA C. *et al.*, "Methodological proposal for evaluation of oil spills environmental vulnerability in rivers", *International Oil Spill Conference Proceedings May 2017*, vol. 2017, no. 1, pp. 1806–1818, 2017.

[KRE 07] KREMER X., Response to Small-Scale Pollution in Ports and Harbours, Operational guide, CEDRE editor, 2007.

[MUT 17] MUTTIN F., CAMPBELL R., OUANSAFI A. *et al.*, "Numerical modelling and experimentation of oil-spill curtain booms: application to a harbor", *Mathematics in Engineering Science and Aerospace*, vol. 8, no. 3, pp. 275–291, 2017.

2

Oil Spill Containment in Open Areas: Four Atlantic and Mediterranean Experiments

The opening feature of a marine area is conducive to a large-scale dispersion of oil spill when its containment fails. Located on the East Atlantic coast of the UK, Spain and Portugal, and the South Mediterranean coast of Spain, the experiments here conducted in open areas are generally closer to harbors where bunkering operations occur or a refuge zone is foreseen. The open characteristic of these areas complicates oil spill containment due to the presence of a sea current generated by a river current along shipping routes. Existing contingency plans can be improved by using model results in parallel to regular drills and training.

2.1. Introduction

In this chapter, four case studies describe marine pollution experiments using floating barriers in the Western Mediterranean Sea and North-east Atlantic Ocean. The experiments were conducted in the ports of Falmouth (UK), Lisbon (Portugal), Puebla del Caramiñal and Algeciras (Spain). Guidelines are available for boom installation in

Chapter written by Frédéric MUTTIN and Rose CAMPBELL.

ports and rivers [DAG 12]. Different booming strategies, defined by various geometries, were used in the experiments (Table 2.1). Three strategies were tested in the case of the Lisbon oil terminal: the first strategy aimed to contain oil from a leaky ship using a quadrangular boom installation; the second strategy carried out in Lisbon involved placing a static boom placed between the terminal and the shore crossing the Tagus River; the third strategy was deploying a dynamic oil recovery in Lisbon by using a chevron towed by two ships. In the Palmonés River estuary, a static boom partially made up of a shore seal boom deployed on the south sandy bank was used. Nautical means were necessary to deploy the booms from ships during the exercises in Falmouth and Ria de Arousa.

Experiment location	Plan geometry	Method of work	Environment	Objective(s)
Falmouth Harbor	Chevron	Static	Coastal	Diversion Concentration Recovery
Lisbon oil terminal Option 1 Option 2 Option 3	 Quadrangle Line Chevron	 Static Static Dynamic	Estuarine Port	Circling Diversion Recovery
Galicia Ria de Arousa	Line	Static	Coastal	Containment
Andalucía Palmonés River estuary	Line	Static Shore seal	River	Diversion

Table 2.1. *Working details of boom installation in open areas*

Three water types were proposed: coastal, estuarine and river. The working objectives of the described boom plans corresponded to the expected effects on the oil pollutant. We define diversion as the deviation of oil from its drifting trajectory, its concentration via actions which increase the oil slick thickness and recovery as the removal of pollutants from the water. Circling reduces the degree of freedom of the oil displacement, and containment protects a sensitive place from the drifting oil.

2.2. Falmouth Harbor

Located inside the Fal River estuary in Cornwall (UK), Falmouth Harbor provides bunkering services. Its oil storage capacity is 50,000 tons and annual fuel oil bunkering in vessels is 750,000 tons. The need for regular training and exercises has been highlighted in a recent event in which a leak occurred on the afternoon of June 10th, 2015, at Falmouth dock, which required the use of booms.

The scenario of the exercise described here was an oil spill starting at 05:30 am on May 12th, 2014, in the Fal estuary in front of the harbor quays. The amount of the pollutant was approximately 4 tons of bunker fuel oil.

The exercise was named MAIA and was centered on a nine-day boom deployment from May 12th, 2014 to May 23rd, 2014. The boom geometry was a chevron, with a downstream head allowing oil diversion and concentration at the chevron center. Oil recovery was simulated using a skimmer and a barge, as shown in Figure 2.1. The total boom plan length was 280 m using 17 curtain sections of 20 m.

A central collection area was created at the head of the chevron, bound by small lengths of the boom. This collector needs a boom overlapping it and exists near the chevron head. Its function is to reinforce the isolation of the pollutant where the oil is accumulating (Figure 2.1). Each boom was inflatable and made up of polyurethane material with a section height of 0.75 m. A total of 19 people were involved in the exercise, including the harbor authority, a tier 2 spill responder subcontractor and the local bunker operator.

Figure 2.1. *Barge for oil recovery and boom curtain for oil containment. MAIA exercise in Falmouth Harbor on May 2014 shows the chevron at the south side and the central oil collector zone at the head of the chevron. Photo by FHC and EIGSI*

The original component of this plan was the use of a set of mooring lines on both sides or arms of the chevron. For each side, a single granite mooring block was placed on the seabed and shackled to the different mooring lines. Each mooring line had a specific length so that the chevron retained its shape, despite varying currents and wind speeds (Figure 2.2).

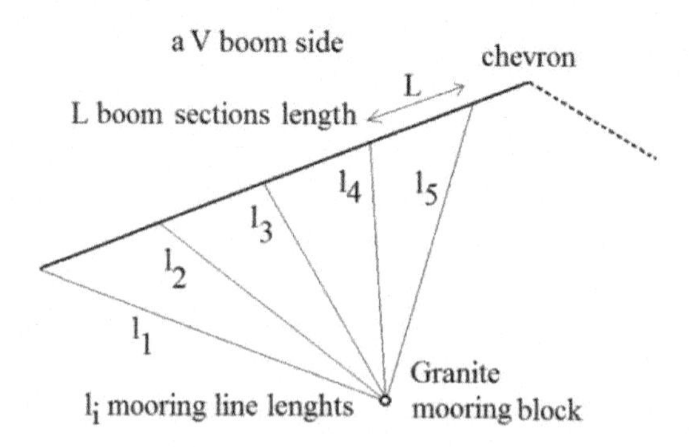

Figure 2.2. *Left side of a V boom indicating mooring line lengths, boom section length and mooring block position*

A GPS positioning tool was used on a catamaran workboat for mooring laying and for deploying the two granite blocks in the positions defined in the contingency plan. The block positions showed an accuracy of 0.3 m. A total of six GPS trackers were set up on the ends of the boom to record their movements for every 30 minutes. A data buoy measured meteorological and tidal conditions at the head of the chevron. A correlation between the boom movement, wind speed and direction and tidal flow was expected [CAS 06].

The most complicated part of the exercise was the installation of moorings, which was carried out a week before the deployment of the booms. On May 12th, two load shackles and data loggers were also deployed on moorings with connections to riser buoys equipped with antennas to transmit the load measurements. Finally, the booms were attached to red floating buoys connected to mooring lays on May 13th, 2014.

More precisely, the load shackles were installed between the mooring blocks and the chains, constituting the bottom connection between each mooring line (Figure 2.3). Tension data was recorded from the load shackles from May 14th at 13:10 to May 15th at 09:00, during which the sensor was disconnected.

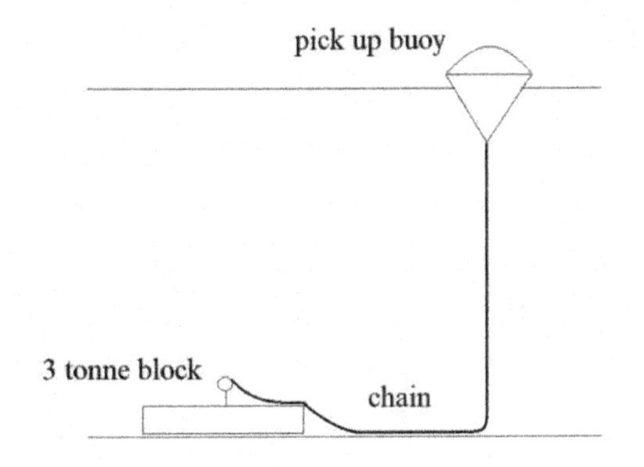

Figure 2.3. *Granite three tonne block, chain and pick-up buoy constituting the oil boom's mooring*

The boom and moorings were retrieved on May 23rd by the catamaran workboat (Figure 2.4) and a motor boat from the harbor. During dismantling, each boom section was unscrewed from the other section at sea, and then dragged into the catamaran, deflated and stored in large storage bags. The moorings were retrieved by the catamaran crane and stored onshore. All of the equipment was rinsed with freshwater.

Training exercises are crucial for a harbor authority responsible for environmental protection. The first reason is knowledge acquisition and management. The activities to be done during an emergency must be clearly defined, prepared for, and included in detailed, illustrated, centralized contingency plans. These plans must include clear instructions about the safe use of equipment and time management during an emergency, including actions that should be avoided.

Figure 2.4. *Moorings deployed via Pendennis catamaran and GPS software. Photo by FHC*

In the next section, we will continue the investigation by looking at an oil terminal in another European country, which is also directly open to the North-east Atlantic Ocean.

2.3. Lisbon oil terminal

The oil terminal of Lisbon Harbor is located in the Tagus River estuary. In such an environment, fast currents can compromise the effectiveness of floating booms [HAN 01]. To highlight the oil spill risk in the Tagus estuary, we can mention the grounding of a 274 m tanker in the bay of Cascais, near Lisbon, on October 17th, 2015, at the vicinity of a breakwater of the marina. The tanker was not carrying crude oil and had only 30 tons of fuel oil. No pollution was observed during this event.

In this section, we will describe an exercise that took place on October 14th, 2014 and October 15th, 2014. To handle the oceanic and atmospheric conditions during this exercise, both numerical and experimental approaches are used [FER 17]. A finite-volume numerical model was used to calculate the water current flow in a series of up to four nested grids surrounding the exercise area and the Tagus estuary [FRA 14]. A set of three drifting buoys equipped with GPS trackers and transmitters allowed Lagrangian tracking of surface water particles. Meteorological forecasting was consulted for the days of the exercise during the preparation stage.

Some details of the environmental data necessary for the response preparedness are given as follows. The numerical computations of the water flow used a 3D domain composed of a set of horizontal grids named sigma layers. The spatial resolution ranged from 10 km on the coarse grid to 100 m on the finest one. The GPS drifting buoys (reference MD-03) used the GSM network to transmit their positions every 15 minutes. Three buoys were attached to the boom to track its position during the exercise. For safety reasons, 48 hours before the exercise, the metocean conditions were checked.

The dominant wind directions at the period of the exercise were in decreasing order of probability: North-East, South-West and North-North-West. Wave direction height and period were computed by using the Swan model in terms of the cardinal wind directions by considering an average wind speed of 10 m.s^{-1}. The tide height ranged from 1.1 m to 3.3 m during the exercise.

As a result of the strong currents in the Lisbon oil terminal, several booming strategies were used during the exercise on October 14th, 2014 and October 15th, 2014 (Figure 2.5). The properties of the curtain boom used are presented in Table 2.2.

Property	October 14th, 2014 and October 15th, 2014
Model reference	NOFI 350 EP
Constitutive materials	Polyurethane and polyvinyl chloride
Total inflated height (m)	0.85
Skirt height (m)	0.5
Float diameter (m)	0.35
Curvilinear mass (kg.m^{-1})	4.7
Chains break (kN)	40 (4 tons)

Table 2.2. *Oil spill curtain boom properties*

The first option of the exercise was to deploy a circling boom. The plan indicates a quadrangular boom geometry to be placed around a ship causing pollution, for example during an oil loading or unloading operation. This response is valuable when the current velocity is less than 0.75 knot. This circling method is shown in Figure 2.5 (Option 1). The quadrangle was replaced by a loop geometry for the boom curtain (Figure 2.6) to save time by reducing the number of anchors.

Figure 2.5. *Lisbon oil terminal with proposition of three boom plans. For a color version of this figure, see www.iste.co.uk/muttin/oil.zip*

Figure 2.6. *Boom installation during the Lisbon oil terminal experiment on October 15th, 2014. Photo by IST-MARETEC*

The second option of the exercise was to install a curtain boom along a line crossing the river from the shoreline to an oil terminal dock. This deployment created a 45° angle between the water current and the boom direction (Figure 2.5, Option 2A-2B), which was ideal for current speeds of up to 1 knot. The boom was deployed at 9:30 am on October 14th and moored at 10:00 am on 2A and 2B bollards following the position labeled option 2, as shown in Figure 2.5. The objective was to divert oil for its recovery on the shore when the current flowed from the right to the left, as shown in the figure.

Figure 2.7 shows a picture of the boom using option 2. At 10:20 am, the boom was trapped in an eddy and pushed against the quays.

Figure 2.7. *Boom moored inside the Lisbon oil terminal using the option 2 plan for spill diversion on October 14th, 2014. Photo by IST-MARETEC*

The third option of the exercise was to simulate dynamic recovery using a chevron geometry (Figure 2.8). The device was towed by two ships in front of the oil terminal. This method was applied to ensure pollutant recovery for most current speeds, as the ships could adjust their speeds according to environmental conditions.

Figure 2.8. *Boom towed near the Lisbon oil terminal during high sea currents for spill recovery. Photo by IST-MARETEC*

Let us quantify the workforce and boats deployed during the entire exercise. The activities involved a total of 20 people, including six researchers of the Lisbon University, and numerous responders, eight from the Lisbon port authority APL, four from the petroleum company and two from the fire brigade. Four APL boats were used during the exercise: a workboat, a pilot boat, a semi-rigid boat and a crane boat. A fire brigade workboat was also involved.

The next section investigates another site in the Iberian Peninsula located further north of Lisbon along the Atlantic coast.

2.4. Galicia Ria de Arousa

An exercise took place in the Ria de Arousa of the Galician region to simulate a place of refuge located on the Atlantic coast of Spain. The North-west coast of Spain is located near the Cap Finisterre where a two-lane traffic pattern is imposed. Located at the confluence of the Atlantic Ocean and the Bay of Biscay, the sea state can present a crossing swell that may cause structural damage to the hull of a ship [ROS 05].

Many pollution response exercises were conducted in the Galicia region, particularly with the support of the European projects ARCOPOL Platform and NETMAR. In 2014, several exercises including tabletop and modeling approaches were organized in different municipalities, ports and regional parks as well as in the Galician Islands National Park.

The exercise presented here focuses on boom deployment from a ship to protect a mussel farm. The pollution scenario is a damaged ship at the quay of a harbor, which is chosen as a place of refuge. The exercise began at 6 am on October 1st, 2014, with a virtual ship asking for assistance after structural damage due to dangerous sea conditions.

At 8 am, the damaged ship was at the Puebla del Caramiñal port, and oil was observed leaking from its hull. The scenario supposes an IFO 180 fuel oil. At 9 am, the authorities were gathered to deploy anti-pollution equipment. At 10 am, an ocean curtain boom from the Galician maritime authority was installed for the prolongation of the port defense (Figure 2.9).

Figure 2.9. *Boom deployment during coastal pollution response operations at A Pobra do Caramiñal Harbor on October 1st, 2014. Photo by INTECMAR, ARCOPOL Platform project, activity 3.3.2. Exercise in Galicia by CETMAR, Consellería do Medio Rural e do Mar (CMRM) – Xunta de Galicia, University of Vigo and activity 5.5.2. Testing in Galicia EIGSI*

As shown in Figure 2.9, the boom curtain rolled off the ship slipway with a stretched geometry. At its other extremity, the boom was moored on an anchor. A boom that becomes stretched is a critical situation. Any ship movement, in particular as a result of its inertia, can overstress the chain and damage the boom or its anchoring points. The tension break of the mooring chains was 50 kN (5 tons). A structural analysis is necessary to predict boom and mooring behaviors. The nonlinear dynamic analysis can be based on a cable structure.

Baixas Rias, Muros Noia, Arousa, Pontevedra and Vigo are an important part of the natural heritage of the Galician coast. As these sheltered places are subject to moderate currents [RAM 13] and are nutrient rich, conditions are favorable for mussel farming. A primary aspect of the Galician contingency plan is the social impact of an oil spill. As a result, the exercise scenario aims to protect the mussel rafts at the vicinity of the pollution source (Figure 2.10). The exercise took place during an ebbing tide with a high water height of 3.10 m at 9:21 am and low water height of 1.20 m at 3:35 pm. The prediction and observation of oil drifting from the port entrance as well as the deployment of other response materials for oil containment and recovery are not detailed.

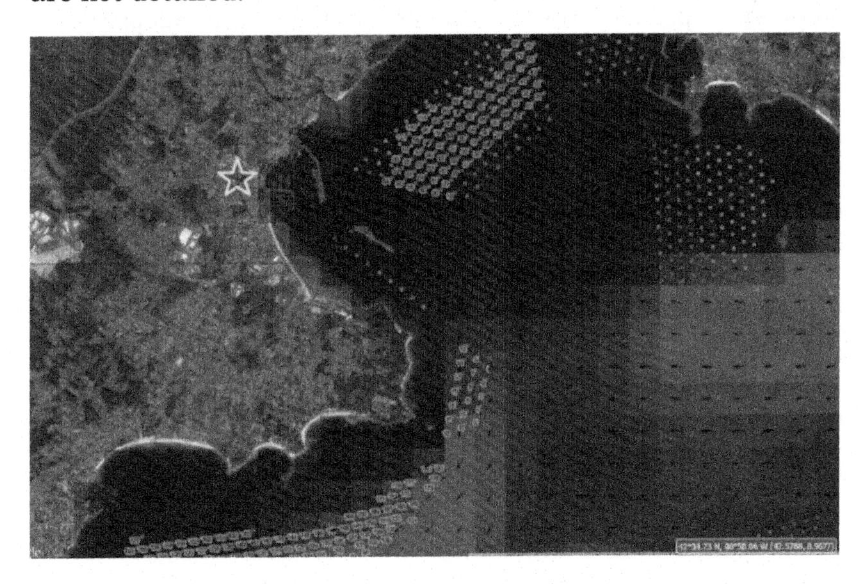

Figure 2.10. *The exercise represents a leaking ship at a refuge place near the harbor entrance. The port of Palmonés is indicated by a yellow star; mussel parks are indicated by red and green circles. The boom will be moored near the red curve. Current directions are indicated by black arrows. For a color version of this figure, see www.iste.co.uk/muttin/oil.zip*

To record in real time the boom geometry during a three-hour exercise, five GPS buoys with GSM communication were fixed on the top leach of the boom curtain (Figure 2.11). These buoys referenced MD-03 were mostly placed in a current flow to describe the oil dispersion mechanism on the sea surface which could be used for calibration of the oil spill model. The uncertainty of the buoy position measurement admitted was generally less than 10 m, whereas the relative positions of the buoys were well known because they were attached to the boom.

In this study, the recorded positions were used for assimilation in the boom structural analysis model. These positions set the moving boundary conditions of the boom section end-points. Thus, we can construct the time-dependent boom displacement with a static model.

Figure 2.11. *GPS marker on the floating barrier. Photo by INTECMAR on behalf of the ARCOPOL Platform project*

Oil spill booms can be modeled by an elastic curvilinear domain on the sea surface. The boom connects its end-points which are supposed to be immobile and respect Dirichlet boundary conditions. The static nonlinear cable equation of a one-dimensional elastic domain can be written as:

$$\frac{d}{ds}T(s) = p.n(s) + q.t(s) \qquad [2.1]$$

where s is the curvilinear coordinate of the domain, T is the boom tension, p is the summation of the hydrodynamic and atmospheric pressures, q is the tangential friction drag and n and t are respectively the unit normal and tangent vectors to the boom. Generally, the drag force q is considered for the simulation of boom towing operations at large speeds, and here it is neglected.

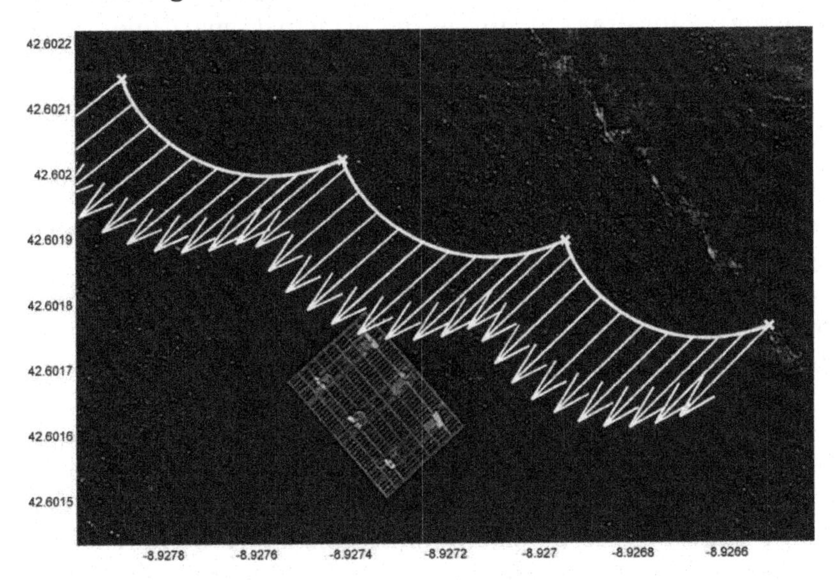

Figure 2.12. *Map view of the computed 2D boom shape with hydrodynamic and atmospheric resultant pressure directions at 10 am on the day of the exercise (October 1st, 2014). For a color version of this figure, see www.iste.co.uk/muttin/oil.zip*

Figure 2.12 shows the result for a three-section cable model on the day of the exercise at 10 am. The hydrodynamic model outputs are available with a 300 m spatial resolution and a 1 hour temporal resolution. At 10 am, the current velocity was 0.15 $m.s^{-1}$, the wind velocity was 4.39 $m.s^{-1}$ and the maximal mooring tension was 18 kg.

The European inter-regional ARCOPOL Platform project was involved in this exercise. The following section will consider another exercise undertaken with this project and belonging to another region of Spain in the Mediterranean Sea.

2.5. Andalucía Palmonés River estuary

A large chemical and hydrocarbon exercise took place in the north west of the Algeciras Bay near the Palmonés River estuary. Gibraltar Strait is a principal maritime route between the Atlantic Ocean and Alboran Sea. Consequently, bunkering activity is high in the Andalusia region, similar to Falmouth harbor in the Cornwall region.

The Algeciras Bay seabed is composed of a canyon in its middle surrounding shallow waters that allow easy ship moorings in a sheltered place. Over the last decades, many ship accidents have resulted in oil pollution. The MV New Flame sank at Europa Point on August 13th, 2007, and the MV Fedra ran aground on October 10th, 2008.

At the back of the bay, twin estuaries of Palmonés and Guadarranque are situated near sandy beaches, terminals, industrial zones, municipal areas and artificial rocky banks. More precisely, Palmonés River has a shallow estuary with a maximum depth of 3.5 m where several environmental studies have been conducted on chronic chemical pollution [CLA 99].

The exercise was organized with public and private organizations in the Palmonés River estuary in the south of Spain in the Bay of Algeciras on February 26th, 2015 (Figure 2.13). It focused on general information exchange, training activities and collaboration at local and provincial levels. Practical courses included testing equipment, responders' training and the use of oceanographic modeling systems.

Figure 2.13. *Aerial picture of the Palmonés River estuary, in the Bay of Algeciras, Spain. The natural protected zone is shown to the south of the picture. The boom placed during the exercise tests the contingency plan of the site in the event of pollution coming from the bay with east wind forcing. The Palmonés city is shown to the north of the picture (Photo (1:25000) by the Consejerías de Obras Públicas y Transporte, Agricultura y Pesca, y Medio Ambiente, Junta de Andalucía)*

A boom was deployed in the Palmonés estuary between a sandy beach to the south and a rocky artificial bank to the north (Figure 2.14). The boom mooring was composed of a shore seal boom ending on a beach and a mooring rope pulled by several responders in a Palmonés city park area. Handling a stressed rope justifies the use of a mechanical model to predict the mooring tension to ensure the safety of operators.

Figure 2.14. *Boom deployment exercise in the Palmonés estuary on February 26th, 2015. Photo by Noticias de la Villa, Consejería de Justicia e Interior, Junta de Andalucía, ARCOPOL Platform project, activity 3.3.1. Exercise in Andalusia Consejería de Agricultura, Pesca y Medio Ambiente and activity 5.5.1. Testing in Andalusia EIGSI*

The boom can be subjected to several forcings in the east-west oriented estuary (Figure 2.15). Wind can blow from the east to the west, resulting in an oil slick drifting upriver. The tidal effect can strengthen the river flow during ebbing. The flooding tide pushes polluted sea water into the Palmonés River estuary. The factors that favor oil spill drift from the sea into the estuarine protected area are large wind coming from the east, waves and flooding tide. On the contrary, river flow, wind blowing from the west and ebbing tide flush pollutants from the river into the sea. We advocate that during an exercise or emergency, the boom must be launched under the calm conditions of current, tide and wind. During east forcing, the boom works in terms of a containment or deviation of a pollutant coming from the sea and going into the river. During west forcing, the boom can be stressed by heavy mechanical pressure due to augmented river flow [MUT 15].

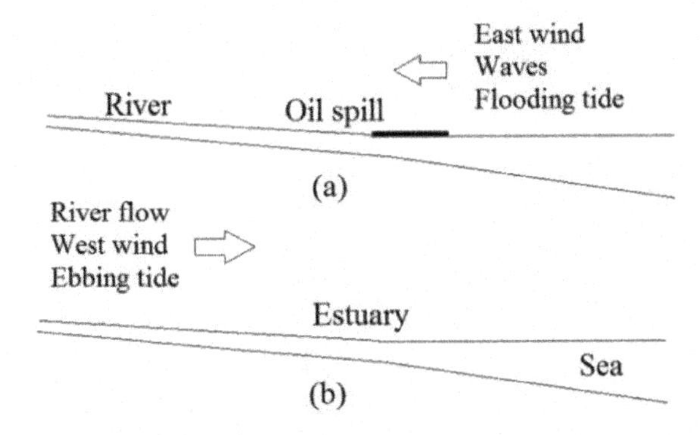

Figure 2.15. *East and west forcing factors on oil spill drift in Palmonés River estuary: (a) east wind, waves and flooding tide; (b) river flow, west wind and ebbing tide*

To force the boom mechanical model during the morning of the exercise, results from the hydrodynamic model and satellite-deduced winds were used. The principal resolutions of these models are given in Table 2.3.

Model	Spatial resolution (m)	Time resolution (h)
NEMO (MyOcean)Hydrodynamic current	~3,000	1
IFREMER CERSAT (MyOcean) Satellite-observed winds	1,000	6

Table 2.3. *Metocean model properties for the horizontal components U and V of the sea current and wind velocities*

A boom length of 70 m was computed using 2D and 3D mechanical models. Figure 2.16 shows a 2D boom geometry prediction that considers a surface current flow resulting from both river discharge and wind. The principal river discharge rates were observed on the Mediterranean coast of the Andalusia region by a monitoring system composed of several automatic hydrographic stations.

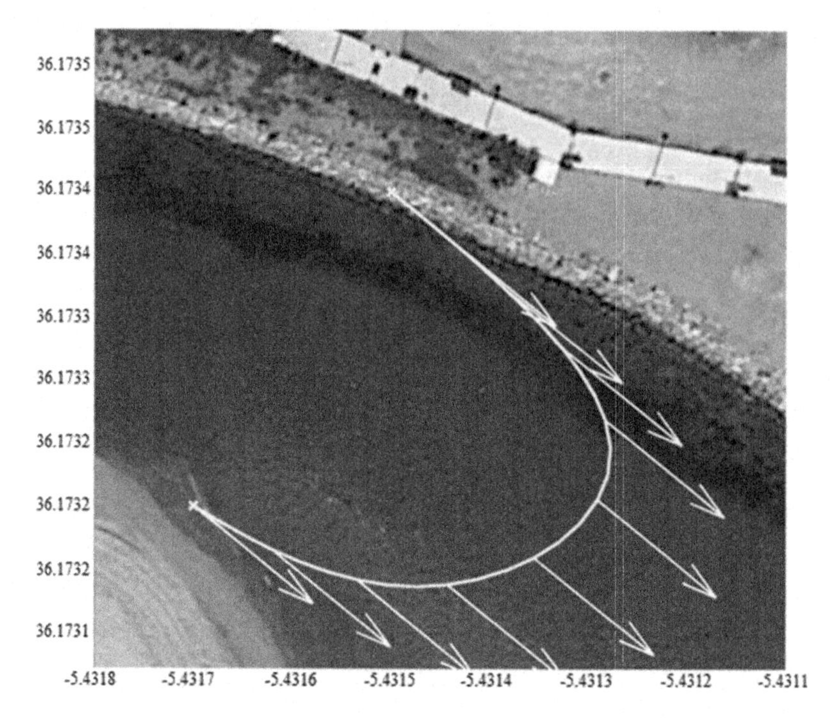

Figure 2.16. *Map view of the simulated 2D barrier. The white arrows indicate the water surface current speed and direction. For a color version of this figure, see www.iste.co.uk/muttin/oil.zip*

Table 2.4 gives the principal results of the 3D model for the boom curtain placed in the estuary. We observed that the mooring tension was low as a result of the low amplitudes of the current and wind.

Parameter	February 26th, 2015 (morning)
Current velocity (m.s^{-1})	0.013
Wind velocity (m.s^{-1})	6.5
Maximal mooring tension (kg)	1
Maximal skirt vertical angle (°)	1.8
Entrainment failure risk (Y/N)	NO

Table 2.4. *Parameters of the 3D model for the boom curtain during the experiment*

We note that, at the first order, the forcing on the boom was proportional to the square of the current velocity. For a river current ten times faster than that observed here (0.13 $m.s^{-1}$), the tension on the mooring line reached 100 kg. In this case, the boom mooring line may not be safely towed by hand. A safety coefficient generally admitted to Navy operations is a factor of 7 for mechanical stress.

2.6. Discussion

The vicinity of a major oil terminal where the bunkering operation occurs is a common point of the experiments deployed in the Falmouth and Lisbon harbors and Algeciras Bay. As a result, local contingency plans already exist in these areas. On the contrary, as the Ria de Arousa is not a major bunkering zone, its protection is described as a part of a regional contingency plan. This open area is defined as a place of refuge.

The main factors of the experiments are given in Table 2.5.

Experiment location	Number of people involved	Length of boom (m)	Maximal flow velocity ($m.s^{-1}$)	Failure of boom plan
Falmouth Harbor	19	280	0.3	NO
Lisbon oil terminal	20	150	2.0	Static Partially Dynamic NO
Galicia Ria de Arousa	50	210	0.3	NO
Andalucía Palmonés River estuary	70	70	0.1	NO

Table 2.5. *Main factors of boom experiments in open areas*

As a result of the partial success of the two static boom plan options in the Lisbon experiment, where large current flow occurs, dynamic experiments were conducted. The dynamic option was successful and did not fail despite strong current conditions, and it could also be used as a recovery mechanism.

Another common aspect of these experiments was the use of hydrodynamic modeling to predict current velocities. Table 2.6 presents the different models, data acquisition platforms and modelers' organizations.

Experiment location	Hydrodynamic Models	Data platforms	Modelers' organizations
Falmouth Harbor	SELFE3D Delft3D		Environmental Hydraulics Institute University of Cantabria
Lisbon oil terminal	MOHID	MOHID Lagrangian wizard WMS server	University of Lisbon Instituto Superio Técnico MARETEC
Galicia Ria de Arousa		ARCOPOL web viewer Meteo-Galicia	University of Vigo INTECMAR
Andalucía Palmonés River estuary	UCA3D	Ocean UCA-Maps	University of Cadix

Table 2.6. *Hydrodynamic models, data platforms and modelers' organizations enrolled in open areas*

The organization of exercises and experiments is mandatory and must be made regularly. International and national recommendations and rules on pollution preparedness specify the required periodicity and main scope of such exercises [MAR 16]. The appellation shared by the community is to use the term Tier 1 for an experiment corresponding to a small local event, Tier 2 for medium spills and Tier 3 for large spills exceeding the national boundary.

The experiments presented concern boom plans and several of them include tests of other techniques and considerations of pollution scenarios for different pollutants. In Falmouth, a barge and a skimmer were used together in the boom plan for oil recovery. In Lisbon, drifting buoys were used as Lagrangian trackers of the water surface behavior, allowing the validation of the hydrodynamic model. In Galicia, several underwater and aerial drones were tested during the pollution scenario, allowing oil slick detection and drift tracking. In the Andalusia experiment, a chemical accident scenario was used in order to train operators for the response equipment and to take necessary actions in case of hazardous and noxious substance spills.

Several distinctive aspects can be identified from the four experiments. First, the maximal current velocity can be significantly different between two open areas. Second, rules and practices are subject to the national legislation applicable in the experiment location. Finally, organization roles and management positions depend on the previous experience and knowledge of local operators, which emphasizes the need for sharing best practices within the oil spill community.

2.7. Conclusion

The experiments presented in this chapter appertain to different European countries and demonstrate similarities and differences in methodology.

The main objective of pollution response experiments is to operate safely, identify uncertainty and test response techniques. The identification of hazards and risks is absolutely necessary for crisis preparedness. We have to consider the known limitations of equipment and use metocean forecasting to support decision-making about boom deployment. Mechanical or artificial unpredictable events can occur during deployment of resources. Targeting pure performances on delay and material optimization may lead to unsafe choices.

The deployment of other response materials is necessary for the emergence of new kinds of potential marine pollutants such as hazardous chemicals and toxic substances. Some general principles can be extrapolated from oil containment by booms such as safety actions, modeling of atmosphere and ocean parameters, and the simulation of techniques set up in the environment.

2.8. Bibliography

[CAS 06] CASTANEDO S., MEDINA R., LOSADA I.J. *et al.*, "The Prestige oil spill in Cantabria (Bay of Biscay). Part I: operational forecasting system for quick response, risk assessment, and protection of natural resources", *Journal of Coastal Research*, vol. 22, no. 6, pp. 1474–1489, 2006.

[CLA 99] CLAVERO V., IZQUIERDO J.J., PALOMO L. *et al.*, "Water management and climate changes increases the phosphorus accumulation in the small shallow estuary of the Palmones River (Southern Spain)", *Science of The Total Environment*, vol. 228, nos 2–3, pp. 193–202, 1999.

[DAG 12] DAGORN L., DUMONT A., Manufactured Spill Response Booms, Operational guide, Centre of Documentation Research and Experimentation on Accidental Water Pollution, Cedre, Brest, 2012.

[FER 17] FERNANDES R., CAMPUZANO F., BRITO D. *et al.*, "Automated system for near-real time prediction of oil spills from Eu satellite-based detection service", *International Oil Spill Conference Proceedings*, vol. 2017, no. 1, pp. 1574–1593, May 2017.

[FRA 14] FRANZ G., PINTO L., ASCIONE I. *et al.*, "Modelling of cohesive sediment dynamics in tidal estuarine systems: case study of Tagus estuary, Portugal", *Estuarine, Coastal and Shelf Science*, vol. 151, pp. 34–44, 2014.

[HAN 01] HANSEN K., COE T.J., Oil spill response in fast currents. A field guide, Report CG-D-01-02, US Coast Guard Research and Development Center Groton CT, October 2001.

[MAR 16] MARITIME AND COASTGUARD AGENCY, Contingency planning for marine pollution preparedness and response: guidelines for ports, OPRC Guidelines for Ports, 3 February 2012, last updated 4 October 2016.

[MUT 15] MUTTIN F., "Structural analysis of oil-spill booms", in EHRHARDT M. (ed.), *Mathematical Modelling and Numerical Simulation of Oil Pollution Problems, The Reacting Atmosphere*, vol. 2, Springer International Publishing, Switzerland, 2015.

[RAM 13] RAMOS V., IGLESIAS G., "Performance assessment of Tidal Stream Turbines: a parametric approach", *Energy Conversion and Management*, vol. 69, pp. 49–57, 2013.

[ROS 05] ROSENTHAL W., LEHNER S., "Results of the MAXWAVE project", *Proceedings of the 14th Aha Hulikoà Winter Workshop*, Honolulu, Hawaii, pp. 1–7, available at: http://www.soest.hawaii.edu/PubServices/2005pdfs/Rosenthal.pdf or https://pdfs.semanticscholar.org/b7db/3e0378119c5e811420f31ba0ec026e66bbcc.pdf, 2005.

3

Polymetallic Pollution in Sentinel Bivalves Across a Semi-open Area: La Rochelle Harbor, France

Currently, the assessment of pollutants' impact on the environment is one of the main objectives worldwide. The second objective is to reach a good chemical and ecological status for marine coastlines. In port areas, marine sediments are often contaminated. They thus pose several problems which have an impact on the marine ecosystem after their immersion in the sea following dredging works.

3.1. Environmental matters of La Rochelle Harbor

La Rochelle Harbor has several areas, including the old harbor and the new harbor "Minimes". The latter is situated in the bay of the city, the provincial administrative center of the Charente-Maritime department, in front of Ré, Oléron and Aix Islands and several kilometers from Charente River estuary. The extension of the "Minimes" basins (Figure 3.1) took place at the cusp of summer 2014.

Chapter written by Marine BREITWIESER, Angélique FONTANAUD and Hélène THOMAS-GUYON.

The aforementioned public work in the harbor increased the capacity for reception by more than 5,000 places. Indeed, La Rochelle Harbor was to be the largest in Europe.

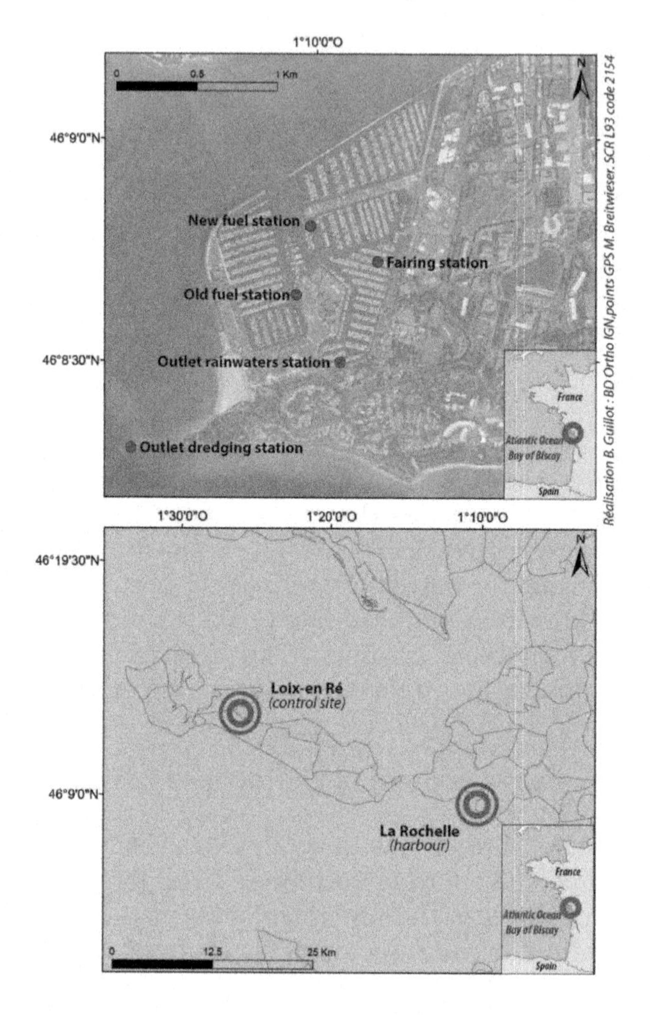

Figure 3.1. *Map of the basin's repartition of La Rochelle Harbor. The red points indicate several study areas: fairing, rainwater outlet and fuel stations (handbook of La Rochelle Harbor) with the extension of Minimes (Lambert 93: EPSG 2154; source: BD Ortho IGN 2014; realization: B. Guillot; GPS data: M. Breitwieser). Site plan of La Rochelle city with the harbor in the Bay of Biscay and reference site, Loix-en-Ré marsh on Ré Island. For a color version of this figure, see www.iste.co.uk/muttin/oil.zip*

The Minimes harbor is composed of several basins for boats at the wharf:

– Lazaret Basin;

– Marillac Basin;

– Bout Blanc Basin;

– Tamaris/New Basin.

All of these areas (Figure 3.1) of the port are subject to different activities. The fairing area is located at the edge of Bout Blanc Basin, the rainwater outlet is situated in Marillac Basin, the old fuel station is in Lazaret Basin and the new fuel station is located near Tamaris and Bout Blanc basins.

3.2. Threats to biodiversity and ecosystem function

The following could be potential sources of heavy metals and contamination:

– copper, zinc and lead are three metals found in the La Rochelle Harbor sediments that exceed quality and quantity guidelines;

– additional examination is required for the contamination in the whole harbor which is above threshold levels;

– historical contamination in La Rochelle Harbor and storm water inflow account for the most part of the impacted sediment;

– benthos, invasive species, fauna and bacterial species undergo changes in highly contaminated areas;

– nearby ecosystems could be disturbed by storms and human activities affecting sediment quality.

According to monitoring networks in harbor areas, four potential sources of chemical contamination have been identified:

– Boat hulls, which are covered by antifouling paint with an uncontrolled release of organostannic compounds. Antifouling paints that contain biocide (heavy metals, copper, tributyltin) are used to prevent algal and microorganism proliferation. Eventually, these paints start disintegrating into the water and sediments of the harbor.

– For an efficient antifouling effect for fairing maintenance, boat hulls need to be sanded and painted every year. Contaminated water from fairing added to rainwater becomes the second contamination source although a filtration treatment is adopted before release into the environment.

– Parking with parked caravans or boater cars can be a source of contamination (hydrocarbons, oils).

– Urban wet-weather effluents, such as runoff waters from roofing, streets and roadways, constitute a high flow rate during storm events and a large amount of pollutants.

The environmental quality of La Rochelle Harbor has been assessed. Previous studies showed a large heavy metals sediment contamination. Indeed, recent works confirm the presence of other contaminant families such as organohaline pesticides and polycyclic aromatic hydrocarbons (PAHs). These recent research observations result in restrictions for commercial fishing and justify recommendations to prohibit consumption of seafood in the harbor [BRE 17].

Systems monitoring the marine environment quality have been put in place and carried out by the *Institut Français de Recherche pour l'Exploitation de la Mer* (IFREMER). These monitoring networks in microbiology (REMI), phytoplankton (REPHY), benthic (REBENT) and chemical contamination

(ROCCH) have been used to respond to the environmental objectives of the European Water Framework Directive.

According to the prefectural order of fairing (prefectural order no. 07-09DISE-DDE, January 1st, 2007) and dredging (prefectural order no. 14EB1000, October 6th, 2014), La Rochelle Harbor must carry out water and sediment analysis, which is controlled by authorities from the *Direction départementale des territoires et de la mer.*

This analysis is carried out on several elements such as physico-chemical (temperature, pH, salinity, turbidity) and bacteriological parameters.

When the harbor areas require dredging as a function of bathymetry, preliminary analysis of organic and inorganic pollutants is carried out on the sediments of basins in need of dredging. The environmental policy of La Rochelle Harbor has been achieved through various actions afloat and ashore, thus allowing the following accreditation:

– since 2006: ISO 14001, the international reference for environmental management;

– since 1985 (from the label creation): Blue Flag, an award for the most committed harbors.

Moreover, staff in charge of environmental problems in La Rochelle Harbor are engaging professionals and sailors in the pursuit of sustainability by cleaning up critical waste with special agents, “the blue brigade”, collecting the bilge waste, black and gray waters, selling organic cleaning products and so on.

Unfortunately, most of the harbors in the world are not the subject of a regular environmental monitoring like this harbor (Figure 3.2).

Figure 3.2. *Staff in charge of environmental problems in La Rochelle Harbor cleaning up critical waste*

3.3. Prevention systems of pollution

This section considers the fairing and fuel station areas in La Rochelle Harbor (Figure 3.3).

3.3.1. *Fairing area*

The average number of fairings per year in La Rochelle Harbor is about 4,700 handlings and includes anti-foulant operations. One of the engaging works of the harbor was to build a settling pond in 2013 with an effective treatment of careening waters (filter system below the fairing area).

3.3.2. *New fuel station*

The storage capacity of gasoline and diesel in the new fuel station of La Rochelle Harbor is 140 m^3 (80 m^3 Go, 25 m^3 SP95, 35 m^3 SP98), with an average distribution of 850 m^3. A few accidental pollution events occurred in a boat fuel station. This particular oil pollution can be treated with dispersants. However, this mixture that is used to move slick on sediments is more toxic than hydrocarbons, as reported by the *Centre de documentation, de recherche et d'expérimentations sur les pollutions accidentelles des eaux* (CEDRE, Brest, France). In order to protect the marine environment from minor pollution, La Rochelle Harbor used biodegrading in order to accelerate the deterioration of pollutants by microorganisms. Furthermore, the new fuel station of the harbor is equipped with retention dispositive features to avoid dripping of fuel during filling of boat tanks.

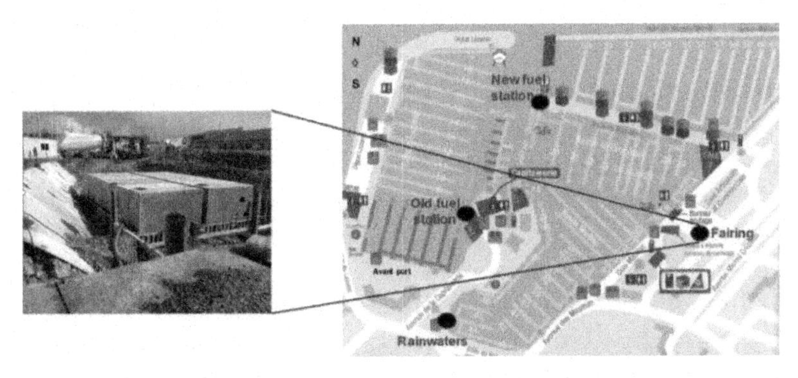

Figure 3.3. *Building process of a settling pond for an effective treatment of careening waters by La Rochelle Harbor. For a color version of this figure, see www.iste.co.uk/muttin/oil.zip*

3.4. Policy regulation and environmental management of La Rochelle Harbor

The environment is listed in the harbor policy regulation of La Rochelle. The settlement was established in article 17 concerning cleanliness of water harbor: "*it is forbidden to use a WC evacuating to the sea in harbor. Any spill waste, unsanitary liquid, and any mineral matter is forbidden and is liable to prosecution*".

This article was created from the Environmental Code (articles L. 210-1 to L. 218-81: seawater and aquatic areas).

The environment department of La Rochelle Harbor adopted (Figure 3.4):

– ISO norm 14001 in 2006: the international system of environmental management that introduces environmental questions about the functioning of a company;

– Blue Flag label in 1985: the first European ecolabel for clean waters and sustainable development.

La Rochelle Harbor signed the charter for a better environment delivered by the A.P.P.A (Atlantic Marinas Association) in 2011.

Recreational yachting is supposed to be the source of a large diversity of non-indigenous species in La Rochelle Harbor. For most of these invaders, few data have been reported. Our study focuses on the pollution effects assessed by means of biochemical biomarkers in three species at selected harbor sites, in order to prove that a multi-biomarker approach detects early signs of impairment in natural bivalve populations and to validate that the standardized procedures of the Water Framework Directive would allow European water quality to be more accurately evaluated.

Figure 3.4. *Rewards for environmental actions of La Rochelle Harbor (Blue Flag, Charter for a Better Environment and International Reference for Eco-Management ISO14001). For a color version of this figure, see www.iste.co.uk/muttin/oil.zip*

3.5. Case study in La Rochelle Harbor using three bivalve species

La Rochelle Harbor shares commitments to improving the protection of birds (with the Royal Society for the Protection of Birds) with local associations and state organizations and the preservation of the marine environment with the Echo-Mer association, in order to develop actions for encouraging people to act responsibly toward the environment. A partnership accord has been signed with the *Ecole de la Mer* (Sea School) association, the purpose of which is to inform, teach and share scientific and technical knowledge of problems such as marine biodiversity and coastal living spaces. Finally, La Rochelle Harbor works with the University of La Rochelle and the French Center for Scientific Research in the Littoral Environment and Societies (LIENSs) laboratory.

This voluntary environmental commitment allowed the signing of a partnership with a research laboratory in order to assess environmental quality in different areas of the marina using three bivalve species, namely oysters, mussels and scallops.

Assessment of the impact of pollutants on marine areas is crucial. It is necessary to obtain a good chemical and ecological status for coastal zones. Several marine species sampling studies carried out in harbors and open areas [MIL 15, LAC 15, LUN 15, BRE 16] have revealed the effect of the risk of diffuse and chronic metallic and organic pollution on oysters (*Crassostrea gigas*), mussels (*Mytilus edulis*) and scallops (*Mimachlamys varia*).

The objective of the project is the environmental management of La Rochelle Harbor. The natural and anthropogenic phenomena affecting the functioning of the harbor must be better understood. The first environmental effort made by La Rochelle Harbor was to improve the wastewater treatment of the fairing area (UV, coal and sand filters). Our study provides the basis for a future biomonitoring program to assess the merits of these environmental quality efforts. Previous studies conducted worldwide (Egypt, Australia, France, Canada and Ivory Coast) have shown that harbor sediments and seawater contain significant proportions of enriched heavy metals with a major biomonitoring requirement [HUS 14, WAL 14, CAR 15, BAK 15, PIP 16]. The pollutant-mediated generation of reactive oxygen species (ROS) in the exposed organisms leads to antioxidant defense mechanisms, which prevents oxidative damage to cellular antioxidant enzymes. Antioxidant enzyme activities in the species of mussels, oysters and scallops have been used as pollutant biomarkers [LAM 09, LUN 10, BUS 05]. The biomarkers used to study the exposure to heavy metals [LUN 15, CAM 13] include the following: laccase (immunomodulation), malondialdehyde (effector of lipid peroxidation), glutathione S-transferase (GST) and superoxide dismutase (SOD, effector of antioxidant defense).

Bivalves were removed from various intra-harbor stations in January 2015, in order to quantify:

– inorganic contaminant concentrations in sessile marine organisms;

– heavily contaminated intra-harbor areas;

– indicator species.

Trace elements and biomarker responses in the digestive glands of bivalves varied between contaminated and reference sites. Energy metabolism is generally slow during winter. We compared four sites (fairing area, rainwater area, new fuel and old fuel stations) affected by anthropogenic activities. Furthermore, a comparison was made with a reference site with much lower levels of heavy metal pollution (Ré Island) [BUS 05, MIL 15].

3.5.1. *Heavy metal concentrations in sentinel bivalves*

Trace elements found in the three species are given in Table 3.1. For oysters (*C. gigas*):

– Cu concentrations were particularly different between Ré Island (reference) and all the other sites;

– Cu and Zn concentrations significantly differed ($p < 0.05$) between the harbor areas (fairing and rainwater areas and both fuel stations) and Ré Island;

– Ag concentrations only differed between the rainwater area and Ré Island;

– Cd concentrations were higher in La Rochelle Harbor (all the sites) than in Ré Island;

– other trace elements did not reveal any principal variations in concentration between the reference and contaminated sites.

	C. gigas				
	Reference Loix en Ré	Rainwater outlet	Fairing area	Fuel area (now)	Fuel area (old)
As	34.22 ± 1.5	28.87 ± 1.2	26.73 ± 1.35	24.78 ± 1.7	28.38 ± 1.92
Cd	2.18 ± 0.19	4.71 ± 0.52	4.28 ± 0.39	3.79 ± 0.39	4.59 ± 0.50
Co	0.47 ± 0.04	0.52 ± 0.06	0.51 ± 0.02	0.41 ± 0.02	0.52 ± 0.04
Cr	0.48 ± 0.07	0.60 ± 1	0.60 ± 0.08	0.41 ± 0.08	0.50 ± 0.06
Cu	100 ± 12^{b}	4208 ± 410^{a}	4442 ± 436^{a}	2090 ± 390^{a}	3207 ± 278^{a}
Fe	439.45 ± 66	468.29 ± 54	464.10 ± 46	354.33 ± 72	364.66 ± 42
Mn	20.68 ± 2.3	44.065 ± 14	22.72 ± 2.24	27.04 ± 10.6	23.03 ± 3.88
Ni	0.893 ± 0.28	0.55 ± 0.16	0.53 ± 0.07	0.41 ± 0.1	0.66 ± 0.13
Pb	1.63 ± 0.11	2.12 ± 0.15	2.19 ± 0.2	1.65 ± 0.27	2.05 ± 0.16
Se	6.31 ± 0.38	5.61 ± 0.42	6.69 ± 0.41	5.94 ± 0.78	6.31 ± 0.68
V	1.14 ± 0.10	1.53 ± 0.25	1.33 ± 0.13	1.34 ± 0.13	1.64 ± 0.21
Zn	1485 ± 260^{b}	6637 ± 785^{a}	7228 ± 539^{a}	4163 ± 715^{a}	4066 ± 314^{a}
Ag	6.52 ± 0.57^{b}	5.64 ± 0.51	4.33 ± 0.56	7.57 ± 1.54	8.92 ± 0.77
Sn	NA	0.10 ± 0.02	0.11 ± 0.02	0.071 ± 0.01	0.15 ± 0.05

	M. edulis				
	Reference Loix en Ré	Rainwater outlet	Fairing area	Fuel area (now)	Fuel area (old)
As	34.25 ± 1.38	21.20 ± 0.38	31.07 ± 0.95	22.32 ± 1.41	22.86 ± 1.52
Cd	0.84 ± 0.06	1.50 ± 0.08	1.13 ± 0.08	0.75 ± 0.09	1.63 ± 0.06
Co	1.82 ± 0.12^{b}	1.06 ± 0.19^{b}	2.09 ± 0.16^{a}	0.92 ± 0.16^{b}	1.45 ± 0.21^{b}
Cr	3.36 ± 0.30	3.15 ± 0.66	3.70 ± 0.38	2.68 ± 0.99	2.97 ± 0.45
Cu	12.95 ± 0.5^{b}	81.71 ± 8.17^{a}	96.67 ± 17^{a}	23.01 ± 4.12^{a}	19.63 ± 2.31^{b}
Fe	2089 ± 205	1404 ± 464	1828 ± 237	925 ± 197	1698 ± 371
Mn	51 ± 4^{b}	100 ± 30^{a}	57 ± 7^{b}	28 ± 4^{a}	48 ± 7^{b}
Ni	7.80 ± 0.51^{b}	3.15 ± 0.70^{a}	7.09 ± 0.44^{b}	3.65 ± 0.61^{b}	4.90 ± 0.60^{b}
Pb	4.44 ± 0.36^{b}	6.42 ± 0.88^{a}	4.60 ± 0.51^{b}	2.57 ± 0.41^{a}	7.49 ± 2.00^{a}
Se	6.97 ± 0.27^{b}	9.52 ± 0.52^{b}	9.51 ± 0.23^{b}	7.18 ± 0.86^{b}	7.27 ± 0.70^{b}
V	4.57 ± 0.43	3.87 ± 1.04	4.22 ± 0.47	2.43 ± 0.37	3.77 ± 0.41
Zn	79 ± 4^{b}	268 ± 18^{a}	111 ± 2^{a}	84 ± 5^{b}	174 ± 70^{a}
Ag	0.13 ± 0.02	0.14 ± 0.02	0.21 ± 0.04	0.07 ± 0.01	0.06 ± 0.00
Sn	0.08 ± 0.01	0.09 ± 0.02	NA	0.06 ± 0.01	0.08 ± 0.01

	M. varia				
	Reference Loix en Ré	Rainwater outlet	Fairing area	Fuel area (now)	Fuel area (old)
As	19.33 ± 0.95	19.04 ± 1.60	16.72 ± 1.45	17.55 ± 1.67	13.73 ± 1.01
Cd	39.85 ± 6.23^{b}	11.85 ± 2.34^{a}	14.34 ± 1.61^{a}	16.60 ± 2.41^{a}	10.77 ± 2.17^{a}
Co	1.51 ± 0.13	1.81 ± 0.22	1.90 ± 0.32	1.37 ± 0.19	1.23 ± 0.17
Cr	1.59 ± 0.18	3.51 ± 0.66	2.67 ± 0.37	2.37 ± 0.35	1.98 ± 0.25
Cu	41 ± 7^{b}	556 ± 61^{a}	503 ± 88^{a}	218 ± 36^{a}	194 ± 35^{a}
Fe	694 ± 72^{b}	923 ± 171^{a}	878 ± 127^{a}	1174 ± 211^{a}	919 ± 230^{a}
Mn	15.76 ± 1.36^{b}	38.95 ± 9.6^{a}	23.68 ± 4.16^{a}	45.84 ± 12.75^{a}	23.34 ± 3.34^{a}
Ni	2.99 ± 0.41	3.67 ± 0.56	3.53 ± 0.58	3.82 ± 0.74	2.80 ± 0.46
Pb	2.20 ± 0.14	2.15 ± 0.34	2.07 ± 0.29	2.68 ± 0.43	2.16 ± 0.26
Se	10.58 ± 0.38	12.39 ± 1.23	10.24 ± 0.71	11.64 ± 1.18	10.05 ± 1.22
V	3.13 ± 0.23	4.49 ± 0.7	4.67 ± 0.75	3.78 ± 0.55	4.20 ± 0.52
Zn	114 ± 23^{b}	165 ± 31^{a}	102 ± 10^{b}	120 ± 18^{b}	74 ± 8^{a}
Ag	10.52 ± 1.03	9.48 ± 1.26	6.69 ± 1.13	6.29 ± 1.23	4.60 ± 0.81
Sn	0.04 ± 0.00	0.15 ± 0.02	0.29 ± 0.04	0.09 ± 0.01	0.17 ± 0.02

Table 3.1. *Trace element concentrations (µg/g dry weight) in the digestive glands of C. gigas, M. edulis and M. varia (n = 10 per site), sampled from the five areas: Loix-en-Ré (reference site), rainwater, fairing and fuel station (new and old) areas. Different superscript letters (a, b) reveal a significant difference ($p < 0.05$)*

For mussels (*M. edulis*):

– Co concentrations were significantly higher in the reference site than in fairing areas;

– Pb concentrations were remarkably higher in rainwater and old fuel areas, but lower in the new fuel area compared with Ré Island;

– Cu concentrations were higher in the harbor (fairing, rainwater and fuel areas) than in the reference site;

– Ni and Mn concentrations were significantly lower in rainwater and old fuel stations;

– Mn and Se concentrations were mainly higher in rainwater and dredging outlet areas.

For scallops (*M. varia*):

– Mn and Cu concentrations were significantly different between the reference site and all the other areas, with the fairing area being the highest contamination area and Ré Island being the lowest contamination site;

– Fe concentrations were mostly higher in all the sites of the harbor than in the reference site;

– Zn concentrations were remarkably lower in the old fuel area and significantly higher in the rainwater area compared with the reference site;

– Cd concentrations were mainly higher in all the areas of the harbor than in the reference site.

M. varia species showed higher modulations of trace elements compared with oysters and mussels.

To study the impact of human activities on coastal waters, bioaccumulated contaminants should be used as valuable tracers of habitat quality in order to monitor the living species (Biodiversity French Agency).

Research on the bioaccumulation capacity of mollusks has been conducted under contamination-controlled conditions. In this study, we use the *in situ* sampling of bivalves in order to:

– share and improve research results on marine areas, species and usages;

– preserve and rehabilitate environmental and ecological functions;

– extend the "land and sea" junction;

– encourage and make progress in professional fishing, aquaculture and shellfish, respecting marine ecosystems;

– flourish industrial and port activities as well as leisure activities, maintaining balanced marine ecosystems;

– arouse passion for sea life in the minds of the large public and involve everyone in the conservation of marine and littoral habitats.

The results from our study on oysters are consistent with the hypothesis about environmental contamination (Cu, Zn, Ag, Cd) from roadworks, roofing panels and especially their alloying agents.

Mussels (*M. edulis*) showed higher concentrations of Cu (rainwater outlet), Co (fairing area) and Zn (rainwater outlet and fuel station), as reported previously [TAC 08]. These results on bivalves reveal a potential risk for consumers' health [MET 08]. Scallops (*M. varia*) showed higher concentrations of Cu, Fe and Mn in harbor sites (fueling station) than in the reference site (Loix-en-Ré county). A potential heavy metal source is antifouling paints used for recovering boat surfaces [BUS 05]. When in excess, these concentrations become toxic.

It is important to compare the habitats of the three bivalves:

– mussels and oysters in the water column filter are contaminated with suspended particles;

– scallops living at the benthic water–sediment interface are affected by contaminated sediments. Therefore, they are more likely to be intoxicated than the other two bivalves.

Measurements of trace elements in sediments are usually carried out by La Rochelle Harbor. For two harbor stations, a high bioavailability of some inorganic contaminants has been found in sediments (see Table 3.2).

	Rainwater outlet	**Fairing area**
Al	20.9	20.9
As	18.0	18.8
Cd	0.14	0.12
Cr	34.3	36.4
Cu	27.5	34.0
Ni	19.8	20.8
P	631	669
Pb	35.1	33.9
Zn	106	113
Hg	0.17	0.11

Table 3.2. *Usual measurements of trace element concentrations (µg/g dry weight) in the sediments sampled from the rainwater outlet and fairing areas in La Rochelle Harbor in 2015*

Several findings have established organic and inorganic contamination of aquatic fauna and the marine ecosystem from dredging and sediment suspension in the harbor [STE 07, MAG 08, JAN 10, SUN 10, KOL 16, CHA 17, BRE 17].

This suggests the need for regular monitoring of La Rochelle Harbor to assess the potential biological effects of this contamination through toxic responses of enzyme activities.

3.5.2. *Application of biochemical biomarkers in monitoring harbor pollution*

Disparities among four biomarkers were observed in the digestive glands of the three species under study: oysters, mussels and scallops (Figure 3.5).

The SOD biomarker was analyzed during an early stage of oxidative stress and before the beginning of cellular damage [VAL 06].

SOD levels were *higher* for:

– mussels sampled from the dredging outlet zone and old fuel station, compared with the reference site;

– scallops collected in Ré Island.

SOD levels were *lower* for oysters and scallops collected from the old fuel station and the fairing area, compared with the reference site.

Compared with bivalves sampled from the reference site Loix-en-Ré, specific SOD activity was different in the three species collected from the harbor areas (fairing, old and new fuel stations).

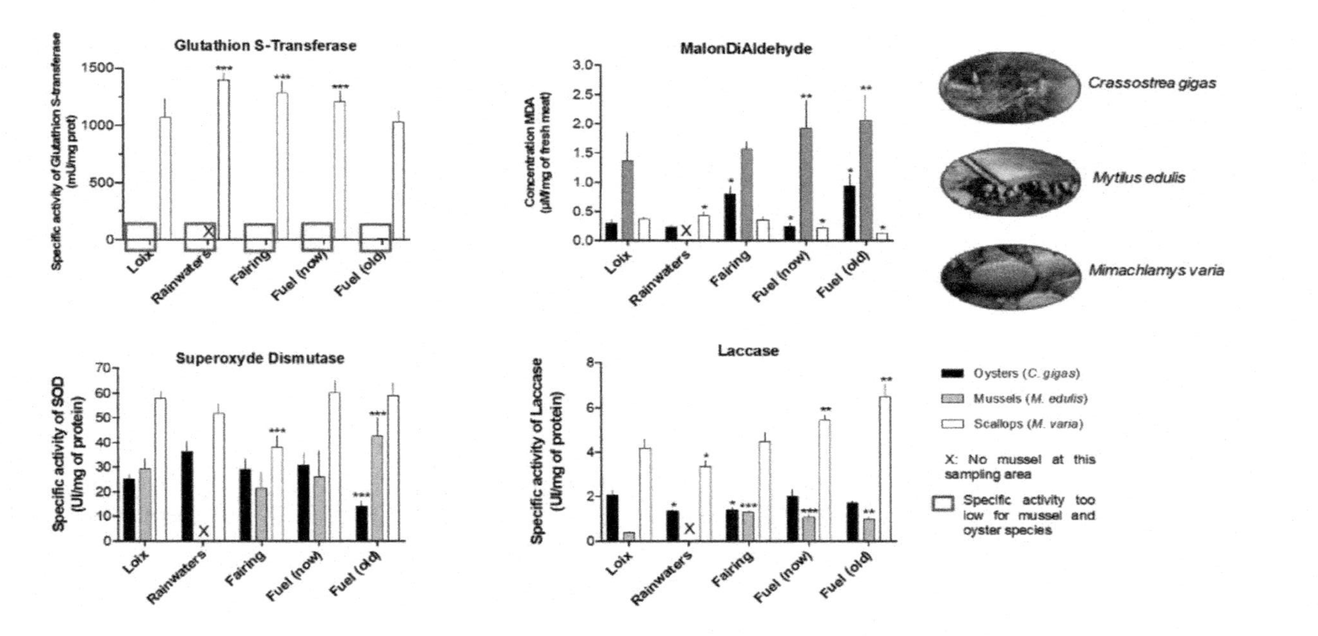

Figure 3.5. *Biomarker levels in the digestive glands of C. gigas, M. edulis and M. varia sampled from a reference site (Loix-en-Ré) and four contaminated sites (rainwater outlet, fairing, new and old fuel stations) in January 2015. Data are presented as mean ± SEM (n = 10). Significant differences were observed between the contaminated sites and the reference site: *p < 0.05; **p < 0.01; ***p < 0.001*

SOD enzyme activity was significantly lower for scallops (*M. varia*) in the fairing area than those in the reference site. This modulation can be attributed to the impact of abiotic factors on heavy metal concentrations in tissues, which may interfere with essential elements and their substitution. As previously mentioned, this species lives in a supra-benthic layer near mudflats and potentially polluted sediments, which can cause a strong reaction against this enzyme involved in the xenobiotic detoxification process. The GST enzyme delivers higher cellular oxidant systems [BOR 05].

In scallops (*M. varia*), lipid peroxidation was analyzed with MDA (malondialdehyde) biomarker concentrations. The analysis showed that lipid peroxidation was activated by MDA in the digestive gland cell membrane:

– for *C. gigas*, all of the samples collected from rainwater, fairing and fuel (old and new) stations contained significantly higher MDA concentrations than those collected in the reference site;

– for *M. edulis*, a significantly higher MDA concentration was found in mussels sampled from both fuel stations;

– for *M. varia*, a significantly lower MDA concentration was evident in the old fuel station compared with Ré Island.

MDA concentrations were significantly lower in scallops from the old fuel station than those in the reference site.

Using this species, MDA concentration can be used as a better biomarker for acute and large contamination than that for small and chronic chemical pollution. Therefore, lipid peroxidation is the last line of defense. Before a potential protein impact, phospholipids will capture a large number of reactive oxygen species (ROS).

The results of the laccase measurement showed that bivalves are sensitive to harbor pollution, which can alter their immune function. The specific activity of laccase was negatively affected in oysters collected from the rainwater outlet and fairing areas. This can be attributed to abiotic factors such as water and air temperatures. This species lives high up in the piling dock harbor. In contrast, the specific activity of laccase was significantly modulated in the digestive glands of *M. varia* collected from the dredging outlet, fairing and fuel (old and new) stations. For *M. edulis*, the specific activity of laccase was significantly lower in the dredging outlet area. For mussels, it was considerably higher in fairing and fuel stations than in the reference site.

A modulation resulting from a mixture of environmental contaminants, hydrocarbons and pesticides generally induces enzyme activities. As a result, they increase the production of reactive oxygen species [MIL 15, BRE 16].

Differences in the vertical positions of oysters, mussels and scallops in the foreshore affect their air exposure time, which thus leads to variations in biomarker activity or content. The sampling date and several abiotic factors, such as salinity, dissolved oxygen, nutrients and chlorophyll, play an important role and thus should be recorded for future study.

3.5.3. *Significance and application of biomarkers in monitoring harbor and environmental pollution*

Oxidative stress biomarkers were measured in the three species (*C. gigas*, *M. edulis* and *M. varia*) sampled from the four contaminated sites of La Rochelle Harbor and from the reference site (Loix-en-Ré).

Biomarker responses (SOD, GST, MDA and laccase) were well correlated with pollutant concentrations in the digestive glands of the three species. Oxidative stress modulation can also depend on several inherent abiotic factors (e.g. oxygen level, temperature, salinity, chemical environment). Moreover, the immunological parameters of aquatic organisms can follow a similar trend.

Oysters	SOD	GST	MDA	Laccase
As	–	–	–	–
Cd	<0.01	–	<0.05	–
Co	–	–	–	–
Cr	–	–	–	–
Cu	<0.01	–	<0.01	–
Fe	–	–	–	<0.01
Mn	<0.01	–	–	–
Ni	–	–	–	–
Pb	<0.05	–	–	<0.01
Se	–	–	–	–
V	–	–	–	–
Zn	–	–	<0.01	–
Ag	–	–	<0.01	<0.01
Sn	<0.01	–	<0.05	–

Mussels	SOD	GST	MDA	Laccase
As	–	<0.05	–	–
Cd	–	–	–	–
Co	–	–	–	<0.05
Cr	–	–	–	–
Cu	–	–	–	<0.01
Fe	–	–	–	<0.01
Mn	–	–	–	–
Ni	–	–	–	<0.01
Pb	–	–	–	–
Se	–	–	–	–
V	–	–	–	<0.05
Zn	–	–	–	–
Ag	–	–	<0.05	–
Sn	–	–	–	–

Scallops	SOD	GST	MDA	Laccase
As	–	–	<0.05	–
Cd	<0.01	–	–	–
Co	–	–	–	–
Cr	–	–	–	–
Cu	<0.01	<0.05	–	–
Fe	–	–	–	–
Mn	–	–	–	–
Ni	–	–	–	–
Pb	<0.01	–	–	–
Se	–	–	–	–
V	–	–	–	–
Zn	–	<0.05	<0.05	<0.05
Ag	<0.01	–	<0.05	<0.01
Sn	<0.01	–	–	–

Table 3.3. *Biomarker responses to heavy metal concentrations in the digestive glands of oysters, mussels and scallops (n = 10 per site and per species) sampled from five areas: Loix-en-Ré and four harbor areas (rainwater, fairing and fuel (new and old) stations); $p < 0.05$, $p < 0.01$ and $p < 0.001$ indicate a significant correlation, and "–" indicates no significant correlation. GST: glutathione S-transferase; SOD: superoxide dismutase; MDA: malondialdehyde*

In scallops, *M. varia* was the most sensitive species showing pollution modulation of the four enzymes (see Table 3.3). The correlations of Ag, As, Cu, Zn, Cd, Pb and Sn with biomarkers (SOD, GST, MDA and laccase) show that mussels and oysters are not always the best bioindicators among invertebrates.

The results indicate that oxidative stress could be induced by:

– the bioaccumulation of oligoelements;

– the water pollution of the La Rochelle Harbor.

Although correlations exist between biomarkers and inorganic contaminants, it is important to consider other *in situ* parameters and abiotic factors (e.g. dissolved oxygen, pH, salinity, temperature) which can modulate oxidative stress and generate cell death in aquatic organisms [LUS 11]. Salinity can activate ROS protein elimination by a hydric potential modulation. Behavioral growth, reproduction and survival rates could be the most impacted by chronic exposure. Indeed, the health and reproduction of bivalve population and communities of harbor species could be altered.

3.6. Conclusion

Unfortunately, few studies focus on the impact of heavy metals on bivalve living in harbor areas. The present study shows that the concentration of heavy metals in harbors is of major concern: inorganic pollutant concentrations in marine mollusks can cause damages. Heavy metals should be assessed regularly in the tissues of marine species. The different functions of the organs are affected by past exposure (water and food). Moreover, the ecological balance of populations in harbor areas is affected by chronic environmental pollution. We suggest to environmental authorities and managers that they estimate the potential toxicity of metal-contaminated effluents and sources. The discharge point must be carefully monitored. Finally, biochemical and cellular biomarker processes could be useful and efficient for quality monitoring in harbor.

From the perspectives of these studies, that aim to develop normalized tools through a biomarker approach, it would be necessary to perform a metallothionein assay during an annual biomonitoring of La Rochelle Harbor for 3 years.

Moreover, to standardize these results in the harbor area, a mathematical model will be developed to help restore the local player's marine environments to sound ecological condition (Europe Strategies 2020).

Environmental protection and harbor user awareness (ISO 14001 Management environmental certification, La Rochelle Harbor) have provided resources to manage this area. A fairing water treatment system with a new fueling station was installed 2 years ago. Both devices and installations currently permit the management to reduce the contaminant concentrations, particularly hydrocarbon leaks and drips. For optimal results and restoration of ecological conditions in harbor areas, it will be appropriate to implement these new treatment processes in all harbors used as commercial and fishing harbors.

3.7. Bibliography

[BAK 15] BAKARY I.A., YAO K.M.A., ETCHIAN O.A.B. *et al.*, "Zinc, copper, cadmium, and lead concentrations in water, sediment, and *Anadara senilis* in a tropical estuary", *Environmental Monitoring and Assessment*, vol. 187, no. 12, pp. 1–11, 2015.

[BIS 10] BISHOP M.J., COLEMAN M.A., KELAHER B.P., "Cross-habitat impacts of species decline: response of estuarine sediment communities to changing detrital resources", *Oecologia*, vol. 163, no. 2, pp. 517–525, 2010.

[BOR 05] BORKOVIĆ S.S., ŠAPONJIĆ J.S., PAVLOVIĆ S.Z. *et al.*, "The activity of antioxidant defence enzymes in the mussel *Mytilus galloprovincialis* from the Adriatic Sea", *Comparative Biochemistry and Physiology Part C: Toxicology & Pharmacology*, vol. 141, no. 4, pp. 366–374, 2005.

[BRE 16] BREITWIESER M., VIRICEL A., GRABER M. *et al.*, "Short-term and long-term biological effects of chronic chemical contamination on natural populations of a marine bivalve", *PLoS ONE*, vol. 11, no. 3, pp. e0150184, 2016.

[BRE 17] BREITWIESER M., VIRICEL A., CHURLAUD C. *et al.*, "First data on three bivalve species exposed to an intra-harbour polymetallic contamination (La Rochelle, France)", *Comparative Biochemistry and Physiology, Part C*, vol. 199C, pp. 28–37, 2017.

[BUS 05] BUSTAMANTE P., MIRAMAND P., "Evaluation of the variegated scallop *Chlamys varia* as a biomonitor of temporal trends of Cd, Cu, and Zn in the field", *Environmental Pollution*, vol. 138, no. 1, pp. 109–120, 2005.

[CAM 13] CAMPILLO J.A., ALBENTAOSA M., VALDÈS N.A. *et al.*, "Impact assessment of agricultural inputs into a Mediterranean coastal lagoon (Mar Menor, SE Spain) on transplanted clams (*Ruditapes decussatus*) by biochemical and physiological responses", *Aquatic Toxicology*, vol. 142–143, pp. 365–379, 2013.

[CAR 15] CARO A., CHEREAU G., BRIANT N. *et al.*, "Contrasted responses of *Ruditapes decussatus* (filter and deposit feeding) and *Loripes lacteus* (symbiotic) exposed to polymetallic contamination (Port-Camargue, France)", *Science of the Total Environment*, vol. 505, pp. 526–534, 2015.

[CHA 17] CHAN J.T.K., LEUNG H.M., YUE P.Y.K. *et al.*, "Combined effects of land reclamation, channel dredging upon the bioavailable concentration of polycyclic aromatic hydrocarbons (PAHs) in Victoria Harbour sediment, Hong Kong", *Marine Pollution Bulletin*, vol. 114, no. 1, pp. 587–591, 2017.

[HUS 14] HUSSEIN A., KHALED A., "Determination of metals in tuna species and bivalves from Alexandria, Egypt", *Egyptian Journal of Aquatic Research*, vol. 40, no. 1, pp. 9–17, 2014.

[KOL 16] KOLAREVIĆ S., KRAČUN-KOLAREVIĆ M., KOSTIĆ J. *et al.*, "Assessment of the genotoxic potential along the Danube River by application of the comet assay on haemocytes of freshwater mussels: the joint Danube survey", *Science of the Total Environment*, vol. 540, pp. 377–385, 2016.

[LAC 15] LACROIX C., RICHARD G., SEGUINEAU C. *et al.*, "Active and passive biomonitoring suggest metabolic adaptation in blue mussels (*Mytilus* spp.) chronically exposed to a moderate contamination in Brest harbor (France)", *Aquatic Toxicology*, vol. 162, pp. 126–137, 2015.

[LAM 09] LAM P.K.S., "Use of biomarkers in environmental monitoring", *Ocean & Coastal Management*, vol. 52, no. 7, pp. 348–354, 2009.

[LUN 10] LUNA-ACOSTA A., ROSENFELD E., AMARI M. *et al.*, "First evidence of laccase activity in the Pacific oyster *Crassostrea gigas*", *Fish & Shellfish Immunology*, vol. 28, no. 4, pp. 719–726, 2010.

[LUN 15] LUNA-ACOSTA A., BUSTAMANTE P., BUDZINSKI H. *et al.*, "Persistent organic pollutants in a marine bivalve on the Marennes–Oléron Bay and the Gironde Estuary (French Atlantic Coast)-Part 2: Potential biological effects", *Science of the Total Environment*, vol. 514, pp. 511–522, 2015.

[LUS 11] LUSHCHAK V.I., "Environmentally induced oxidative stress in aquatic animals", *Aquatic Toxicology*, vol. 101, no. 1, pp. 13–30, 2011.

[MAG 08] MAGNI P., RAJAGOPAL S., VAN DER VELDE G. *et al.*, "Sediment features, macrozoobenthic assemblages and trophic relationships (delta C-13 and delta N-15 analysis) following a dystrophic event with anoxia and sulphide development in the Santa Giusta lagoon (western Sardinia, Italy)", *Marine Pollution Bulletin*, vol. 57, nos 1–5, pp. 125–136, 2008.

[MET 08] METIAN M., BUSTAMANTE P., HÉDOUIN L. *et al.*, "Accumulation of nine metals and one metalloid in the tropical scallop *Comptopallium radula* from coral reefs in New Caledonia", *Environmental Pollution*, vol. 152, no. 3, pp. 543–52, 2008.

[MIL 15] MILINKOVITCH T., BUSTAMANTE P., HUET V. *et al.*, "*In situ* evaluation of oxidative stress and immunological parameters as ecotoxicological biomarkers in a novel sentinel species (*Mimachlamys varia*)", *Aquatic Toxicology*, vol. 161, pp. 170–175, 2015.

[PIP 16] PIPPY B.A., KIDD K.A., MUNKITTRICK K.R. *et al.*, "Use of the Atlantic nut clam (*Nucula proxima*) and catworm (*Nephtys incisa*) in a sentinel species approach for monitoring the health of Bay of Fundy estuaries", *Marine Pollution Bulletin*, vol. 106, nos 1–2, pp. 225–235, 2016.

[STE 07] STEGER K.K., GARDNER J.P., "A laboratory experiments on the effects of variable suspended sediment concentrations on the ecophysiology of the porcelain crab *Petrolisthes elongatus* (Milne Edwards, 1837)", *Journal of Experimental Marine Biology and Ecology*, vol. 344, no. 2, pp. 181–192, 2007.

[SUN 10] SUNDSTEIN C.M., NOHR G.R., KJERULF P.J., "Degradation of mussel (*Mytilus edulis*) fecal pellets released from hanging long-lines upon sinking and after settling at the sediment", *Canadian Journal of Fisheries and Aquatic Sciences*, vol. 67, no. 9, pp. 1376–1387, 2010.

[TAC 08] TACON A.G.J., METIAN M., "Global overview on the use of fish meal and fish oil in industrially compounded aquafeeds: trends and future prospects", *Aquaculture*, vol. 285, no. 1, pp. 146–158, 2008.

[VAL 06] VALAVANIDIS A., VLAHOGIANNI T., DASSENAKIS M. *et al.*, "Molecular biomarkers of oxidative stress in aquatic organisms in relation to toxic environmental pollutants", *Ecotoxicology and Environmental Safety*, vol. 64, no. 2, pp. 178–189, 2006.

[WAL 14] WALKER T.R., MACASKILL D., "Monitoring water quality in Sydney Harbour using blue mussels during remediation of the Sydney Tar Ponds, Nova Scotia, Canada", *Environmental Monitoring and Assessment*, vol. 186, no. 3, pp. 1623–1638, 2014.

4

Oil Spill Dispersant Use: Toxicity on Marine Teleost Fish

The application of dispersants on oil slicks is a controversial oil spill response technique due to the toxicity that it can induce on sub-surface organisms. In this chapter, we summarize the recent findings concerning the toxicity of dispersants on teleost fish. Focus has been placed on the lethal and sub-lethal toxicity induced by dispersed oil and on the future research needs in this field. By compiling and discussing recent and past studies, this chapter is of interest to decision-makers who are aiming to establish a framework for dispersant use policies.

4.1. Introduction

Dispersants are chemicals that are commonly used to clear oil spill (18% of oil spills between 1995 and 2005 [CHA 07]). Spread by boat and/or plane, dispersants decrease the interfacial tension between oil and water, allowing the oil slick on the surface of the sea to break into small droplets in the water column (extensively described in Lessard and Demarco [LES 00]). This process prevents

Chapter written by Thomas MILINKOVITCH, Stéphane LE FLOCH and Hélène THOMAS-GUYON.

damage to organisms dwelling on the sea surface (e.g. seabirds and marine mammals) and enhances petroleum bacterial degradation [CHU 95, SIN 99]. On the contrary, oil droplet formation results in an increase in the total surface area for the partitioning of oil-soluble compounds (from oil to water), leading to an increase in the seawater concentration of potential toxicants [COU 05, MIL 11a, WU 12]. Consequently, several studies showed an increase in PAH (polycyclic aromatic hydrocarbon) concentration in teleost fish due to dispersant use [RAM 04, MIE 05, SCH 09, LEE 11, MIL 11a]. Therefore, in spite of the advantages of this technique for sea surface organisms, dispersion of the petroleum slick is considered to be problematic by decision-makers due to the toxicity of the dispersed oil on sub-surface organisms. In order to assist decision-makers in achieving their aim to draw a framework for dispersant use policies, in this chapter, we summarize the recent findings concerning the toxicity of dispersants on marine teleost fish and discuss the future research needs in this field. Focus has been placed on recent findings considering the fact that these studies involved the use of third-generation dispersants, which are most commonly used [MER 05]. Teleost fish species were chosen as the species of interest due to the environmental and socioeconomic influences that an alteration of these species could induce: they are considered to be key species of ecosystems by determining their trophic structures [GIL 84], and they are involved in fishing economy and food security [FAO 10].

4.2. Lethal toxicity of dispersant application on teleost fish

In recent studies, toxicity of dispersants has been evaluated using their third-generation formulations. Since those dispersants are considered to be intrinsically less toxic [MER 05], focus is on the oil–dispersant mixture, i.e. chemically dispersed oil. Thus, recent experimental approaches compared the toxicities of chemically dispersed oil and undispersed oil. In teleost fish, focus has been placed on the early life stages, considering the high sensitivity to contaminants in these stages [AMI 08]. In this way, some authors evaluated mortality in juvenile teleost fish due to dispersant use [OTI 09, MIL 11a]. However, the majority of the authors focused on earlier life stages such as larvae [COU 05, HEM 11] and embryos [AND 09, SCH 09, LEE 11, VAN 12, WU 12]. Only a few studies showed that both dispersed and undispersed oil have similar toxicities (Hemmer *et al.* [HEM 11] in larvae of *Atherinops affinis*). Conversely, most of the recent studies (cited above) exposed significant and intense differences in mortality of teleost fish with dispersed and undispersed oils, highlighting an increased mortality due to dispersant use. In addition to the fact that toxicity is induced by both contaminants (dispersant and oil), Otitoloju and Popoola [OTI 09] demonstrated in *Clarias gariepinus* that the toxicity of dispersed oil is due to the interaction between oil and dispersant. This finding is in line with the mode of action of dispersant, which increases the partitioning of oil toxic compounds and consequently the exposure of organisms to these compounds (previously described in the Introduction, section 4.1).

Measurements of lethal toxicity increased our knowledge regarding the toxicity of dispersants. However, contamination levels (concentrations and/or exposure periods) that were used in these experimental approaches and that were necessary to induce mortalities were higher than those observed *in situ* (total petroleum hydrocarbon concentrations are rarely higher than 50 mg/L and decrease rapidly after 12 h, according to Reed *et al.* [REE 04]). Thus, complementary to these studies, which measure mortality due to unrealistic levels of contamination, further studies have considered realistic levels of contamination and the induced sub-lethal toxicity.

4.3. Sub-lethal toxicological effects of dispersant application on teleost fish

Sub-lethal toxicity of dispersant application was mostly investigated at sub-organism levels. The presumed toxicity was evaluated through the incorporation of contaminants using bioconcentration measurement as well as bile metabolites (see [DAN 11, MIL 11b, MIL 11c, MIL 12, MIL 13a, MIL 13b, DUS 15a, DUS 15b] for recent studies on this topic). Although these measurements were of great interest by presuming a potential toxicity of dispersant application, they did not reveal its actual toxicity. Biomarkers of hydrocarbon biotransformation, such as ethoxyresorufin-*O*-deethylase (EROD) activity, were mainly studied in the liver of teleost fish, taking into consideration its status as a detoxification organ [VAN 03]. Although EROD is not considered as a biomarker of toxicity, it can be considered as a biomarker of a toxicological process setup, since its activation by PAHs induces the production of ROS (reactive oxygen species) and the consequent deleterious effects [AMI 08]. Studies have revealed an increase of EROD production due to the application of dispersants in *Sebastes schlegeli* and *Gadus morhua* (in Jung *et al.* [JUN 09]; Lee *et al.* [LEE 11]), suggesting deleterious effects and no increase in *Dicentrarchus labrax* (Dussauze *et al.* [DUS 15a, DUS 15b]). Regarding deleterious effects, by searching for histopathological liver lesions in *Siganus canaliculatus*, Agamy [AGA 12] did not reveal any significant difference between dispersed and undispersed oils. The authors suggested that the discrepancy could be due to the low sensitivity of the fish species. Regarding the gills, more histopathological lesions were observed in fish exposed to dispersed oil compared to undispersed oil (in *Epinephelus chlorostigma* by Agamy [AGA 13]), which showed that dispersant use enhances the toxic effects of oil on gills. This finding is in accordance with a study conducted by Mendonça Duarte *et al.* [MEN 10], who highlighted an alteration of the functional integrity (ion regulation) of this organ in

Colossoma macropomum contaminated with chemically dispersed oil.

A study conducted by Milinkovitch *et al.* [MIL 13b] in the heart of *Liza aurata* revealed significant differences in enzymatic antioxidant responses between dispersed and undispersed oils. These results could be linked to the diminished cardiac performances [MIL 13a] and the cardiac abnormalities observed in earlier life stages (in *Atherinops affinis*, Anderson *et al.* [AND 09]) when fish are exposed to dispersed oil.

To conclude, most of the investigations revealed that the application of dispersants enhances the toxicological effects of oil for teleost fish, suggesting the environmental toxicity of this response technique. However, an overview of the corpus of literature incites us to be cautious in our conclusions, since it reveals that further investigations are needed before deciding on the toxicity of dispersant use.

4.4. Future research needs

Regarding the literature, criticisms could be made concerning the significance of previous experimental approaches and the resulting conclusions. Indeed, firstly, most of the studies investigating the toxicity of dispersed oil considered the standardized methods of Singer *et al.* [SIN 99]. However, although this method can be considered as suitable for experimental approaches, it does not take into account the presence of oil droplets as a determinant of toxicity. Recent studies conducted by Olsvik *et al.* [OLS 12]; Milinkovitch *et al.* [MIL 11b], [MIL 11c], [MIL 12], [MIL 13a], [MIL 13b] and Dussauze *et al.* [DUS 15a], [DUS 15b] have considered this parameter, and the results suggest that when the oil is under turbulent processes (such as swell or waves), the application of a dispersant does not enhance the toxicity of oil for teleost fish. Further experimental approaches conducted in a way that best simulates the chemical dispersion of oil (such as the wave tank studies by Li *et al.* [LI 08]) could be of interest.

Secondly, toxicity of dispersants has mainly been studied by exposing teleost fish to water-soluble compounds, and, to the best of our knowledge, available data regarding the toxicity of dispersed oil-contaminated sediments to benthic or supra-benthic fish are scarce. However, dispersant application modulates the availability of PAHs for the sediments and thus the toxicity for the benthic environment [YAM 03]. This sediment contamination highlights another lack of information regarding the chronic toxicity of dispersed oil-contaminated sediments. Indeed, through trophic and direct pathways, sediment hydrocarbon contamination is likely to induce chronic toxicity [OLI 10]. Investigations that take into account long-term effects due to both pelagic and benthic compartments could be of interest to better estimate the toxicity as well as to evaluate the long-term resilience following dispersant application. To this end, field studies, such as those conducted on mangrove ecosystems by Baca *et al.* [BAC 05], seem appropriate.

Finally, most of the investigations evaluating the toxicity of dispersants were conducted at the sub-organism or organism level. However, a few studies were conducted to understand the impairment due to dispersant use at higher levels of biological organization. Although some authors studied individual performances from the perspective of their connection with proxies of Darwinian fitness [KER 12, CLA 13], no study has directly evaluated the fish population impairment induced by dispersed oil. In this way, *in situ* sampling through national observation networks following oil spill could be of great interest.

4.5. Conclusion

In this chapter, we summarized the recent findings concerning toxicity of dispersants in teleost fish. Most of the studies investigating lethal and sub-lethal toxicity suggest environmental impact due to dispersant application. However, we also exposed the limits of these studies by making some criticisms concerning the significance of the experimental approaches used. Indeed, further investigations are recommended to assess the long-term effects of dispersed oil, the impact at population level and the contamination through several compartments of ecosystems (benthic and pelagic). Here, it is of interest to establish dispersant use policies, since it concerns the impairment of dispersed oil on teleost fish and consequently the potential environmental and socioeconomic impacts. However, with regard to both toxicity and the advantages of dispersant use, other representative communities of ecosystems should be considered by oil spill decision-makers.

4.6. Bibliography

[AGA 12] AGAMY E., "Histopathological liver alterations in juvenile rabbit fish (*Siganus canaliculatus*) exposed to light Arabian crude oil, dispersed oil and dispersant", *Ecotoxicology and Environmental Safety*, vol. 75, pp. 171–179, 2012.

[AGA 13] AGAMY E., "Impact of laboratory exposure to light Arabian crude oil, dispersed oil and dispersant on the gills of the juvenile brown spotted grouper (*Epinephelus chlorostigma*): a histopathological study", *Marine Environmental Research*, vol. 86, pp. 46–55, 2013.

[AMI 08] AMIARD-TRIQUET C., AMIARD J.C., *Les biomarqueurs dans l'évaluation de l'état écologique des milieux aquatiques*, Lavoisier, Paris, 2008.

[AND 09] ANDERSON B.S., ARENELLA-PARKERSON D., PHILLIPS B.M. *et al.*, "Preliminary investigation of the effects of dispersed Prudhoe Bay Crude Oil on developing topsmelt embryos, *Atherinops affinis*", *Environmental Pollution*, vol. 157, no. 3, pp. 1058–1061, 2009.

[BAC 05] BACA B., WARD G.A., LANE C.H. *et al.*, "Net environmental benefit analysis (NEBA) of dispersed oil on nearshore tropical ecosystems derived from the 20 year "TROPICS" field study", *International Oil Spill Conference Proceedings*, vol. 2005, no. 1, pp. 1–4, Miami, May 2005.

[CHA 07] CHAPMAN H., PURNELL K., LAW R.J. *et al.*, "The use of chemical dispersants to combat oil spills at sea: a review of practice and research needs in Europe", *Marine Pollution Bulletin*, vol. 54, no. 7, pp. 827–838, 2007.

[CHU 95] CHURCHILL P.F., DUDLEY R.J., CHURCHILL S.A., "Surfactant-enhanced bioremediation", *Waste Management*, vol. 15, nos 5–6, pp. 371–377, 1995.

[CLA 13] CLAIREAUX G., THÉRON M., PRINEAU M. *et al.*, "Effects of oil exposure and dispersant use upon environmental adaptation performance and fitness in the European sea bass, *Dicentrarchus labrax*", *Aquatic Toxicology*, vols 130–131, pp. 160–170, 2013.

[COU 05] COUILLARD C.M., LEE K., LE GARE B. *et al.*, "Effect of dispersant on the composition of the water-accomodated fraction of crude oil and its toxicity to larval marine fish", *Environmental Toxicology and Chemistry*, vol. 24, no. 6, pp. 1496–1504, 2005.

[DAN 11] DANION M., LE FLOCH S., LAMOUR F. *et al.*, "Bioconcentration and immunotoxicity of an experimental oil spill in European seabass (*Dicentrarchus labrax* L.)", *Ecotoxicology and Environmental Safety*, vol. 74, no. 8, pp. 2167–2174, 2011.

[DUA 10] DUARTE R.M., HONDA R.T., VAL A.L., "Acute effects of chemically dispersed crude oil on gill ion regulation, plasma ion levels and haematological parameters in tambaqui (*Colossoma macropomum*)", *Aquatic Toxicology*, vol. 97, no. 2, pp. 134–141, 2010.

[DUS 15a] DUSSAUZE M., DANION M., LE FLOCH S. *et al.*, "Growth and immune system performance to assess the effect of dispersed oil on juvenile seabass (*Dicentrarchus labrax*)", *Ecotoxicology and Environmental Safety*, vol. 120, pp. 215–222, 2015.

[DUS 15b] DUSSAUZE M., DANION M., LE FLOCH S. *et al.*, "Innate immunity and antioxidant systems indifferent tissues of seabass (*Dicentrarchus labrax*) exposed to crude oil dispersed mechanically or chemically with Corexit 9500", *Ecotoxicology and Environmental Safety*, vol. 120, pp. 270–278, 2015.

[FAO 10] FAO FISHERIES AND AQUACULTURE DEPARTMENT, The State of World Fisheries and Aquaculture 2010, Part 1, World Review of Fisheries and Aquaculture, pp. 3–89, available at: http://www.fao.org/3/a-i1820e.pdf, Rome, 2010.

[GIL 84] GILINSKY E., "The role of fish predation and spatial heterogeneity in determining benthic community structure", *Ecology*, vol. 65, no. 2, pp. 455–468, 1984.

[HEM 11] HEMMER M.J., BARRON M.G., GREENE R.M., "Comparative toxicity of eight oil dispersants, Louisiana sweet crude oil (LSC), and chemically dispersed LSC to two aquatic test species", *Environmental Toxicology and Chemistry*, vol. 30, no. 10, pp. 2244–2252, 2011.

[JUN 09] JUNG J.H., YIM U.H., HAN G.M. *et al.*, "Biochemical changes in rockfish, *Sebastes schlegeli*, exposed to dispersed crude oil", *Comparative Biochemistry and Physiology Part C Toxicology & Pharmacology*, vol. 150, no. 2, pp. 218–223, 2009.

[KER 12] KERAMBRUN E., LE FLOCH S., SANCHEZ W. *et al.*, "Responses of juvenile sea bass, *Dicentrarchus labrax*, exposed to acute concentrations of crude oil, as assessed by molecular and physiological biomarkers", *Chemosphere*, vol. 87, no. 7, pp. 692–702, 2012.

[LEE 11] LEE K., KING T., ROBINSON B. *et al.*, "Toxicity effects of chemically-dispersed crude oil on fish", *International Oil Spill Conference Proceedings*, vol. 2011, no. 1, pp. 163–180, Portland, March 2011.

[LES 00] LESSARD R.R., DEMARCO G., "The significance of oil spill dispersants", *Spill Science & Technology Bulletin*, vol. 6, no. 1, pp. 59–68, 2000.

[LI 08] LI Z., LEE K., KEPKAY P. *et al.*, "Wave tank studies on chemical dispersant effectiveness: dispersed oil droplet size distribution", *NATO Science for Peace and Security Series C: Environmental Security*, pp. 143–157, 2008.

[MER 05] MERLIN F.X. (ed.), Traitement aux dispersants des nappes de pétrole en mer, Operational Guide, p. 54, CEDRE, Brest, 2005.

[MIE 05] MIELBRECHT E.E., WOLFE M.F., TJEERDEMA R.S. *et al.*, "Influence of a dispersant on the bioaccumulation of phenanthrene by topsmelt (*Atherinops affinis*)", *Ecotoxicology and Environmental Safety*, vol. 61, no. 1, pp. 44–52, 2005.

[MIL 11a] MILINKOVITCH T., KANAN R., THOMAS-GUYON H. *et al.*, "Effects of dispersed oil exposure on bioaccumulation of polycyclic aromatic hydrocarbons and mortality of juvenile *Liza ramada*", *Science of the Total Environment*, vol. 409, no. 9, pp. 1643–1650, 2011.

[MIL 11b] MILINKOVITCH T., NDIAYE A., SANCHEZ W. *et al.*, "Liver antioxidant and plasma immune responses in juvenile Golden grey mullet (*Liza aurata*) exposed to dispersed crude oil", *Aquatic Toxicology*, vol. 101, no. 1, pp. 155–154, 2011.

[MIL 11c] MILINKOVITCH T., GODEFROY J., THÉRON M. *et al.*, "Toxicity of dispersant application: biomarkers responses in gills of juvenile golden grey mullet (*Liza aurata*)", *Environmental Pollution*, vol. 159, no. 10, pp. 2921–2928, 2011.

[MIL 12] MILINKOVITCH T., LUCAS J., LE FLOCH S. *et al.*, "Effect of dispersed crude oil exposure upon the aerobic metabolic scope in juvenile golden grey mullet (*Liza aurata*)", *Marine Pollution Bulletin*, vol. 64, no. 4, pp. 865–871, 2012.

[MIL 13a] MILINKOVITCH T., THOMAS-GUYON H., LEFRANÇOIS C. *et al.*, "Dispersant use as a response to oil spills: toxicological effects on fish cardiac performance", *Fish Physiology and Biochemistry*, vol. 39, no. 2, pp. 257–262, 2013.

[MIL 13b] MILINKOVITCH T., IMBERT N., SANCHEZ W. *et al.*, "Toxicological effects of crude oil and oil dispersant: biomarkers in the heart of the juvenile golden grey mullet (*Liza aurata*)", *Ecotoxicology and Environmental Safety*, vol. 88, pp. 1–8, 2013.

[OLI 10] OLIVA M., GONZALES DE CANALES M.L., GRAVATO C. *et al.*, "Biochemical effects and polycyclic aromatic hydrocarbons (PAHs) in Senegal sole (*Solea senegalensis*) from a Huelva estuary (SW Spain)", *Ecotoxicology and Environmental Safety*, vol. 73, no. 8, pp. 1842–1851, 2010.

[OLS 12] OLSVIK P.A., LIE A.K.K., NORDTUG T. *et al.*, "Is chemically dispersed oil more toxic to Atlanticcod (*Gadus morhua*) larvae than mechanically dispersed oil? A transcriptional evaluation", *BMC Genomics*, vol. 13, no. 1, pp. 702–703, 2012.

[OTI 09] OTITOLOJU A.A., POPOOLA T.O., "Estimation of "environmentally sensitive" dispersal ratios for chemical dispersants used in crude oil spill control", *The Environmentalist*, vol. 29, no. 4, pp. 371–380, 2009.

[RAM 04] RAMACHANDRAN S.D., HODSON P.V., KHAN C.W. *et al.*, "Oil dispersant increases PAH uptake by fish exposed to crude oil", *Ecotoxicology and Environmental Safety*, vol. 59, no. 3, pp. 300–308, 2004.

[REE 04] REED M., DALING P., LEWIS A. *et al.*, "Modelling of dispersant application to oil spills in shallow coastal waters", *Environmental Modelling & Software*, vol. 19, nos 7–8, pp. 681–690, 2004.

[SCH 09] SCHEIN A., SCOTT J.A., MOS L. *et al.*, "Oil dispersion increases the apparent bioavailability and toxicity of diesel to rainbow trout (*Onchorynchus mykiss*)", *Environmental Toxicology and Chemistry*, vol. 28, no. 3, pp. 595–602, 2009.

[SIN 99] SINGER M.M., AURAND D., BRAGIN G.E. *et al.*, "Effect of dispersants on oil biodegradation under simulated marine conditions", *International Oil Spill Conference Proceedings*, Seattle, vol. 1999, no. 1, pp. 169–176, March 1999.

[VAN 03] VAN DER OOST R., BEYER J., VERMEULEN N.P.E., "Fish bioaccumulation and biomarkers in environmental risk assessment: a review", *Environmental Toxicology and Pharmacology*, vol. 13, no. 2, pp. 57–149, 2003.

[VAN 12] VAN SCOY A.R., ANDERSON B.S., PHILIPS B.M. *et al.*, "NMR-based characterization of the acute metabolic effects of weathered crude and dispersed oil in spawning topsmelt and their embryos", *Ecotoxicology and Environmental Safety*, vol. 78, pp. 99–109, 2012.

[WU 12] WU D., WANG Z., HOLLEBONE B. *et al.*, "Comparative toxicity of four chemically dispersed and undispersed crude oils to rainbow trout embryos", *Environmental Toxicology and Chemistry*, vol. 31, no. 4, pp. 754–765, 2012.

[YAM 03] YAMADA M., TAKADA H., TOYODA K. *et al.*, "Study on the fate of petroleum-derived polycyclic aromatic hydrocarbons (PAHs) and the effect of chemical dispersant using an enclosed ecosystem, mesocosm", *Marine Pollution Bulletin*, vol. 47, nos 1–6, pp. 105–113, 2003.

5

Extreme Environments: The New Exploration/Production Oil Area Problem

Due to their extreme abiotic factors, marine Arctic and deep-sea ecosystems present many similarities. Consequent to the increasing need in petrochemical products, the Arctic and deep sea are subject to a strong interest from oil industry companies. Invasive anthropic activities could lead to a disturbance in these very sensitive ecosystems. Feedback on the potential biological impact of oil industry activities, particularly the behavioral impact of petroleum products and by-products on the biocenosis, is still scarce for these areas. Furthermore, questions about the assumed positive impacts of technical responses, such as *in situ* burning or chemical dispersants, are still debated.

5.1. Introduction

Oil demand associated with the depletion of easily exploitable oil deposits has led the oil industry to intensify the search for new oil areas and to exploit deposits that were

---

Chapter written by Matthieu DUSSAUZE, Stéphane LE FLOCH and Florian LELCHAT.

previously considered as economically unprofitable. These new oil research areas are mostly located in "offshore" areas in deep waters and in the Arctic. The oil and gas industry is massively investing in the European Arctic (from Greenland to the Russian border). Recently, the Italian industry Ente Nazionale Idrocarburi (ENI) has discovered a large hydrocarbon resource at 1,400 m depth off the coast of Egypt, and large reserves are suspected to exist in Cyprus, Israel, Lebanon and the French Guiana Exclusive Economic Zone (EEZ). Thus, an augmentation of human activities in these new exploitation areas will undoubtedly increase the probability of oil spills. Chemical releases, whether chronic or accidental, could upset the balance of these fragile ecosystems whose ecotoxicological resilience is still largely under-documented. In case of an oil spill at sea, the balance costs versus the environmental benefits drive the implemented factors to respond effectively and to limit the impacts of pollution [KOY 04]. Net Environmental Benefits Analysis (NEBA) aims to provide answers which will assist the authorities in an operational emergency context after an oil spill. Several operational methods can be used to limit the effects of such pollution, including *in situ* burning, mechanical recovery or the use of chemical dispersants (hereafter simply called dispersants). Analyses of the International Tanker Owners Pollution Federation (ITOPF) database have shown that dispersants were used in 18% of the 258 marine incidents recorded between 1995 and 2005 [CHA 07]. Among their advantages, they increase the natural dilution of oil and consequently its bioavailability for biodegradation phenomenon; they also decrease the amount of oil slick that washes up on the shore [LES 00]. However, dispersants may increase the toxicity of oil by increasing its bioavailability and therefore exacerbate the effects of oil pollution [CHA 07, WOL 98]. For example, the use of dispersants during an oil spill increases the exposure of fish to petroleum compounds [RAM 04]. In this context, the question of the effects on keystone species of both untreated oil and dispersed oil should be pointed out, and the use of dispersants in the new oil research areas should be carefully considered. Furthermore, the question of the behavior of these chemicals in extreme environments must be focused on.

5.2. Arctic and deep sea: ecosystem specificities

5.2.1. *Characteristics of marine Arctic areas and related zoocenose*

The Arctic marine ecosystem is extremely seasonally marked. Due to its high latitude position, important photoperiod variations will occur during the year and have important ecological consequences. Primary production will occur during a very short period in comparison to tropical or even temperate areas, and will stop during the darker time of winter. Many zooplankton species will enter into diapause during this period to reduce their metabolism in order to survive during this starving time [HIR 96, TAR 16].

Consequently, Arctic biocenosis is typically characterized by specific demographic strategies. Arctic species generally have low reproductive rates, intergenerational times and lifetimes longer than tropical or temperate species [CHA 05].

This extreme environment has also influenced the life trait and ecophysiology of species. For example, fish have evolved to limit the effect of cold water on their metabolism. The internal fluid osmolarity of teleostei is generally of an order of 300–400 mosm.L^{-1}, which corresponds to a freezing point between –0.6 and –0.8°C. However, due to the presence of salt, polar waters exhibit freezing temperatures at –1.8°C. Fishes living in these areas have therefore developed adaptive mechanisms to survive at such temperatures. In order to avoid the formation of ice crystals that would induce cellular necrosis and possibly lead to death, fishes have developed antifreeze compounds that are mainly composed of glycoproteins. These osmolytes, present in their blood, and their interstitial liquids will bring down the freezing point of these fluids [CHA 05].

These highly specialized life characteristics decrease their adaptive potential, which can already be strongly influenced by climatic stress induced by the global change [KOV 11, WAS 11, POS 13]. The trophic chain of marine polar ecosystems and especially sympagic ecosystems (i.e. characterized by the presence of sea ice) can be considered to be relatively short [CHA 05]. In a very schematic way, primary producers (phytoplankton species) will be grazed on by zooplankton, mainly copepods, which will be predated by fishes or other zooplankton. The latter will serve as energy supplies to the following links in the food chain, that is, birds and mammals.

5.2.2. *Characteristics of deep-sea areas and related zoocenose*

Deep-sea ecosystems are still not fully described. In the middle of the 19th Century, it was considered that there was no or very short life at depth below 600 m [AND 06]. This azoic theory was rejected in the second part of the 19th Century [RAM 11]. With the help of a telegraphic cable installed at the bottom of the deep ocean, scientific expeditions demonstrated the presence of several species living in the abyss.

Life in the deep ocean is subject to several extreme abiotic factors. Pelagic area can be divided into two zones. The photic or euphotic zone, from 0 to approximately –200 m (depending on the turbidity of the water), is the area where sunlight (at least 1%) is still perceptible and allows photosynthesis. The aphotic zone is the area where less than 1% of sunlight is present, corresponding to approximately –200 m until the bottom of the ocean (–11,034 m being the deepest in the Mariana trench). Hydrostatic pressure is an important and predictable abiotic factor and increases concomitantly with depth, one more bar of pressure for each 10 m of depth. This abiotic factor can reach more than 1,100 bar of pressure in the deepest areas of the ocean, such as hadal zones which correspond to trench zones.

The global ocean represents more than 70% of the earth's surface. In addition, it should be kept in mind that the average height of the earth's land surface is 2,400 m below the surface of the ocean. *De facto*, the deep ocean is unanimously recognized as the largest and most representative habitat on earth [RAM 10]. So, in theory, deep-sea floor ecosystems can be considered the most representative benthic area of the ocean, whereas shallow and coastal areas as very exotic (and therefore "extreme") from this perspective.

However, the trophic web in the deep sea mainly depends on the highest water layers. Bioavailable energy comes from dissolved organic matters (DOMs) and particulate organic matters (POMs). This marine snow is generated by fecal pellets, exudate and particles of plankton which sink from shallow waters [THI 03]. At the margin of this constant flow, a large quantity of organic matter can fuel the deep-sea floor through the sinking of massive dead animals such as whales. This occasional food supply can provide energy for deep-sea biocenosis for a decade [GAG 92]. Finally, specific deep-sea biological communities have been discovered, with a strong abiotic factor gradient being present everywhere. The most documented case is the profusion of life around deep-sea hydrothermal vents [MCC 15]. However, other complex ecosystems have also been described around cold deep-sea brine pools [OLU 04].

5.2.3. *Impacts of temperature and high pressure on physiology*

Deep-sea and Arctic ecosystems are considered as extreme environments. This is because of high hydrostatic pressures (up to 1,114.58 bars) and low temperatures (4°C corresponding to the maximum density of seawater) in the deep sea and very low temperatures (4°C down to –50°C) in the Arctic.

The temperature has an effect on both energy and volume, whereas the hydrostatic pressure only has a volume effect [PRA 07]. The effect on volume, both at a low temperature and a high hydrostatic pressure, is similar and affects the cellular phospholipid membrane [SOM 92, HAZ 95, MAC 97, PRA 07]. These two abiotic factors will decrease its fluidity [MAC 84]. As reviewed by Brown and Thatje [BRO 14b], according to the membrane composition, an increase in the pressure of 100 MPa corresponds to a decrease in the temperature between 13 and 21°C [SOM 92]. Similarly, an increase of 2.8°C can reverse the consequence of the membrane fluidity at a hydrostatic pressure of 10 MPa [DES 79].

In an adapted zoocenose, the index of fatty acid unsaturation will increase in the phospholipid bilayer in order to maintain a sufficient fluidity of the membrane. Otherwise, cellular ions and molecule transports will not be sufficiently effective and may thus lead to the loss of homeostasis in case of inefficient energy supply [SOM 92]. Beyond deadly issues such as cellular implosion or freezing, for a zoocenose unadapted to cold or high pressure, the consequence can be a decrease in metabolism due to the lowest mitochondrial activity [SOM 92, PEC 02].

Finally, as reported by Tomanek [TOM 10], cellular responses to low temperature and a high hydrostatic pressure can lead to biogeographic limitations. The antagonist effect on membrane and protein of an increase in pressure and a decrease in temperature suggests that both low temperatures and high pressures could lead to the bathymetric limitation of species distribution [BRO 11].

Low temperature and pressure can also affect metabolic rates through conformational changes in protein folding. Nevertheless, the effects of pressure and cold temperature on protein are quite different from those affecting the phospholipid membrane. The most impacted proteins are certainly enzymes which are in charge of the entire metabolism. Thus, piezophiles (organisms living at a high pressure) and psychrophiles (organisms living at a cold temperature) have adapted their metabolic engineering by evolving specific extremozymes (or extremophilic enzymes).

– Pressure

High pressure has a similar effect on protein as high temperature. They affect protein stability and possibly lead to protein unfolding, which can result in the loss of activity [OHM 13]. Notably, piezozymes (or piezophilic enzymes) generally demonstrate a higher thermostability than their mesophilic counterparts, even if they come from a cold-adapted organism [ABE 01]. Beyond general deleterious proteic superstructural constraints, overpressure can also dramatically reduce the volume of the active sites of catalytic domains and inhibit enzymatic reactions [OHM 13]. Adaptive mechanisms comprise the selection of highly stable protein models and the production of osmolytes and/or chaperone proteins [SIM 06].

– Temperature

Cold environments impose thermodynamic challenges on protein and particularly enzymes. Low temperature strongly inhibits enzymatic pathways by lowering chemical reaction rates [ZEC 01]. In a reaction, cold-adapted organisms have evolved psychrozymes (or psychrophilic enzymes). These proteins are characterized by very flexible structures and low stabilities [GOM 04]. Indeed, they generally contain a high amount of α-helix compared to the β-sheets [GIA 01]. Moreover, they exhibit a reaction rate (k_{cat}) that is not affected by low temperatures. This biochemical behavior is the result of the destabilization of the active sites and/or the entire protein which allows a better mobility and flexibility at a low enthalpy state [FEL 03]. As a result, psychrozymes have an average activity that is 10 times more important than mesophilic enzymes but with a constitutive thermolability [GOM 04]. These biochemical features seem to be the fruit of a selective adaptation for low stability enzymes as well as for highly reactive enzymes [FEL 03].

– Psychrozymes versus piezozymes

For an organism living at high pressures and cold temperatures like the deep-sea biocenosis, the efficiency of the metabolic rate must therefore be a consensus between protein stability and catabolic reactivity. However, data on this possible antagonism and resulting biochemical adaptations are still scarce. This “enthalpy paradox” urges for better integrated studies covering all the aspects driving such environments.

5.3. Oil spill in extreme environments

5.3.1. *The Arctic marine environment*

The advent of anthropocene initiated by the development of agriculture, followed by the first industrial revolution in the 19th Century and finally the recent demographic explosion, has affected the climate’s engine functioning. The natural fluctuations of climate have been shaken by human activities. This phenomenon is known as global change [PAR 03]. Today, all scenarios tend toward a worsening of the phenomenon in the next century [OVE 14]. Polar regions appear particularly sensitive to global change [MAC 12]. Indeed, in Arctic regions, a major reduction in ice areas has been observed in the last 30 years [COM 08, PER 08], opening up new shipping routes and making the exploration and exploitation of new promising oil reserves possible [USG 00, PIE 08].

The increase in traffic and human activities in these highly sensitive regions enhances the risk of pollution by petroleum compounds. To respond to oil spills, chemical dispersants are commonly used. These compounds may affect the bioavailability of oil and therefore its potential toxicity [CHA 07]. Thus, their use in Arctic regions is still debated due in part to the lack of knowledge on their effects on organisms. The question of the effects of both oil and chemically dispersed oil on Arctic organisms must be raised.

There are only a few reported cases of significant oil spills in Arctic regions. The rupture of the Komi pipeline in Russia in 1994 [WAR 97] is, along with the sinking of the oil tanker Exxon Valdez, one of the main cases. The latter took place in Alaska in March 1985 at the Prince William Sound. Following this accident, the "Double Hull" amendment was promulgated, requiring vessels built after July 6th, 1996 to be provided with this protection (MARPOL Convention, Rule 13F). From an ecological perspective, Short *et al.* [SHO 03] demonstrated that oil from this sinking was subject to a slow biodegradation and always affected the environment 10 years after the accident. Several experimental studies have focused on hydrocarbon exposures to polar fish. It was found that, for example, the metabolism of the Antarctic fish *Pagothenia borchgrevinki* increases when it is exposed to soluble oil fractions [DAV 92]. Christiansen *et al.* [CHR 10] showed an inverse effect with a decrease in the metabolism of *Boreogadus saida*, the polar cod, exposed to the soluble fraction of North Sea petroleum. Polar cod is a key species in the Arctic ecosystem [ORL 09, HOP 13] and is used as an indicative ecological proxy of environmental effects following contamination with hydrocarbons in the Arctic environment [STA 97]. In addition, for a few years now, polar cod has been recognized as a relevant biological model for studying the effects of hydrocarbons on cold water species [JON 10, NAH 10b, GAR 13]. A study on the growth of polar cod showed that ingestion of petroleum-contaminated food significantly reduced the growth performance of these fish [CHR 95]. Individuals of the same species also exposed to soluble petroleum fraction demonstrate genotoxicity through the measurement of significant DNA damage, even at low concentrations of PAH [NAH 10a].

Indubitably, impacts caused by hydrocarbon pollution on key species such as polar cod or species from a lower trophic level such as Calanoids could disrupt the energy transfer with a strong cascading effect in a different level of the trophic web [HOP 13].

Despite this fact, compared to temperate ecosystems, the potential environmental consequences of an oil spill in the Arctic ecosystem [NAH 10b, OLS 11, GAR 13] and the dispersant as a technical response [DUS 14a] remain under-documented. In an emergency context, this could lead to extrapolating results of studies from temperate species to Arctic species. A study [OLS 11] investigated this question with the determination of acute toxicity tests on 11 Arctic and 6 temperate species exposed to 2-methyl naphthalene. They concluded that values of survival metrics for temperate regions are transferrable to the Arctic for the chemical 2-methyl naphthalene with the consideration of some uncertainties. A comparative work of acute toxicity tests on several species from Arctic to temperate regions concludes that when the test is performed using pure compound: naphthalene for de Hoop *et al.* [DEH 11], 2-methyl-naphthalene for Olsen *et al.* [OLS 11] or the dispersant Corexit 9500A for Hansen *et al.* [HAN 14], no difference in sensitivity is reported between Arctic and temperate fish or invertebrates [BEJ 17].

Nevertheless, the question of the sub-lethal and chronic effects of petroleum compounds is still largely unknown. In this context, a study [DUS 14b] assessed the sub-lethal effects of the same set of dispersed oil on two fish species. They were chosen due to their representative place in their respective ecosystems, polar cod for Arctic areas and seabass (*Dicentrarchus labrax*) for temperate areas. This study showed an interspecific variation and, therefore, demonstrates the strong experimental bias, possibly caused by the use of temperate species to assess the environmental impact of dispersed oil on polar areas. This seminal work highlighted the important need for ecotoxicological investigations in polar areas, particularly for long-term effects of oil pollution.

5.3.2. *The deep sea*

Based on the estimates of the International Energy Agency in "World Energy Outlook 2013" [IEA 13], more than 40% of remaining recoverable resources of conventional oil are located in offshore areas and consequently in areas with important depths. However, it has been reported that, for a usual offshore infrastructure, an incident reported by a company increases by 8.5% for each 30 m supplementary depth [MUE 13]. The available data on the ability of ecosystems to respond to pollution caused by hydrocarbons in the function of depth are still scarce. The example of the *Deepwater Horizon* accident in the Gulf of Mexico in April 2010, one of the biggest oil spills ever reported [JOY 15], was the most striking demonstration. During this blowout, 731 million L of crude oil were released in the environment at a depth of 1,500 m [GRI 12, KUJ 11]. For the first time, dispersant (2.9 million L of COREXIT 9500) was injected directly into the *Macondo* wellhead [KUJ 11] without any environmental consideration for deep-sea ecosystems in an NEBA context (see Introduction). Following this, a large plume of dispersed oil was reported between 1,100 and 1,400 m under the sea surface [CAM 10, HAZ 10]. This oil plume has therefore affected communities in these areas.

Benthic communities [MON 13] and especially deep-sea coral communities [WHI 12] were especially affected and could take decades to recover [MON 13]. Benthic communities are key players of the food web in deep-sea habitats and consequently play a prominent role in carbon cycling. However, during an oil spill, according to NEBA, each of the technical response must be analyzed. However, there is nothing in literature which describes the efficiency of NEBA for an accident such as *Deepwater Horizon*. Due to this lack of data, it is urgent to study the biological impact of oil in deep-sea areas.

Although many experimental studies have been performed to analyze the potential impact of oil on coastal areas [MIL 11, CLA 13, LEF 14, THE 14, TIS 15, DUS 15a, DUS 15b], very few experimental studies have been conducted on the biological impact of oil on deep-sea environments [VEV 09, OLS 16, DUS 16b]. Thus, there is an important necessity, in the NEBA context, to gather a maximum of information from ecotoxicological trials in the deep-sea environment and particularly under a high hydrostatic pressure [MES 13, BRO 14a]. Hydrostatic pressure, like hydrocarbons, is known to have an important effect on biological systems [SOM 91, SÉB 98, THE 00]. To date, a potential synergistic effect between these two parameters has been simply ignored in literature. To fully understand the impact of a pollutant at depth, it is fundamental to couple experimental studies with deep species under a high hydrostatic pressure. However, this parameter is difficult to control and requires very expensive technologies (e.g. hyperbaric chamber). Therefore, very few laboratories are able to produce data on this field. To alleviate this technological constraint, some alternative approaches may be performed. Ecotoxicological studies can be conducted using easy handling organisms like piezotolerant species (deep coral species [DOD 07] or deep amphipods [CAM 05]) at an atmospheric pressure, and with shallow-water species such as shrimp (*Palaemonetes varians*) [COT 12] or demersal fish (*Dicentrarchus labrax*) [DUS 16b] under a high hydrostatic pressure. These ecotoxicological studies can also perform physiological measurements under hydrostatic pressure using isolated cells [DUS 16a]. However, the convenience of laboratory work can never substitute the ideal approach of *in situ* studies.

5.4. Sensitivity of deep sea and polar ecosystems to oil pollution

Increasing amounts of evidence tend to highlight the greater sensitivity of Arctic and deep-sea species to hydrocarbon pollution compared with species from temperate or tropical ecosystems [CHA 05]. Species from deep-sea and Arctic ecosystems can, for the most part, be considered as K strategist [CLA 79, NOR 12]. Their growth can be defined as slow and of a long shelf life, compared to species from the surface or an environment where the temperatures are higher. This is more exhibited in deep environments where most of the trophic supply comes from the surface by sedimentation of organic matter (particulate organic matters, dead animals). The energetic supply can also be limited in a polar area. During winter, the leak from solar light and the sea ice cover conducts to limit the primary production, leading to a reduction in energy transfer from the lowest to the highest level of the food web.

Life in the cold environment can lead to morphological changes. For a similar ecological niche, polar animals are generally larger than those living at the lowest latitudes [DEB 77]. This is demonstrated by the example of the three different species of zooplankton of the genus *Calanus* that live in Arctic waters. *Calanus hyperboreus*, which is found in the very cold water masses of northernmost regions, has an average size of 4.5–6.0 mm and a lipid content of 65% for stage V (the last stage of the copepod phase before becoming a mature individual). *Calanus finmarchicus* can be qualified as a temperate-boreal species, mainly distributed in the sub-Arctic Atlantic waters but more present in the relatively warmer waters where the Gulf Stream still has an impact, will be of an average size of 2.0–3.0 mm and a lipid content of 34% for stage V. *Calanus glacialis*, a true cold-water species, found mainly in Arctic waters, is considered as an ecological key species in the Polar Basin [HAN 12]. It belongs to the calanoid order, which is representative of intermediate water between the icy waters of high latitudes and the warmer waters of the northern Atlantic. It presents an average size of 3.0–4.0 mm and a lipid content of 61% for stage V.

A comparison study of ecotoxicology of Hansen [HAN 13] with *C. glacialis* and *C. finmarchicus* tried to demonstrate that *C. glacialis* is less sensitive than *C. finmarchicus* to an acute exposure of water soluble fraction of diesel. Nevertheless, some variations in the experimental protocol call for caution with the results. For example, the experimentation was performed at 2°C for *C. glacialis*, whereas it was performed at 10°C for *C. finmarchicus*, even if these results support other similar studies (e.g. Hjorth and Nielsen [HJO 11]). Considering the toxicokinetic rule that the elimination rate decreases proportionally with an increase in the size of the tested organism [HAN 13], *C. glacialis* is about four times larger than *C. finmarchicus*. Hence, lipophilic compounds such as PAH accumulated more specially in lipid-rich tissues of the organism due to their high Log Kow [ATS 95]. In this case, the proportion of lipid in their organisms (*C. glacialis* 0.26; *C. finmarchicus* 0.23) has an important role. Indeed, if the PAH concentration in lipid is the same between a lipid-rich and a lipid-poor organism, the total amount of PAH is higher in the lipid-rich than in the lipid-poor organism. Due to difference in kinetics, lipid-rich organisms will take more time to depure these xenobiotics and consequently be exposed to more contaminants than lipid-poor organisms [LAS 90].

Furthermore, animals living in cold environments with a relatively low energy intake will exhibit a slower metabolism than species from temperate or tropical environments [PEC 02]. In theory, the accumulation of contaminants would be slower. In return, if the metabolism of these individuals is slower, then all the mechanisms of protection of the organism (detoxification and depuration) would also be slowed down [CHA 05]. Due to polar winters, Arctic marine life has evolved to survive during "scarcity" times. One of these evolutions is to make important lipid reserves in the aim to maintain homeostasis [LEE 06]. However, potentially contaminant petroleum-based chemicals such as PAHs are lipophilic and will be able to concentrate in fats. As a result, individuals would be affected over long-term and even future generations through the allocation of these energy reserves for reproduction (e.g. gamete maturation).

Consequently, the persistence of petroleum compounds in Arctic and deep-sea environments after a pollution incident is an important problem. Due to the very specific abiotic factors of the ecosystem, the question of the natural biodegradation rate of petroleum compounds must be asked. The long-term contamination or, in the better case, the quick bioremediation of the ecosystem and particularly the benthic ecosystem, depends on the natural biodegradation of chemicals mainly mediated by autochthonous microbiota.

5.5. Behavior of oil in Arctic and deep-sea areas

Directly after an oil spill, the oil slick will be subjected to weathering processes, resulting in physical and chemical changes in oil over a period of a few hours to several years. They are due to mechanisms of evaporation, dissolution, dispersion, emulsification, sedimentation, adsorption on sedimentary particles, photo-oxidations or biodegradation. The intensity of these mechanisms depends on the intrinsic physicochemical properties of the oil as well as abiotic (temperature, photoperiod, wind, wave) and biotic (biodegradation) factors.

The bioavailability of hydrocarbons is paramount in the biodegradation process [LEF 99]. Biodegradability of oil increases with the specific surface of oil droplets generated by a dispersion phenomenon. In theory, the dispersant technical response coupled with the energy of waves and wind will favor the dispersion of oil. By increasing the oil-water exchange area, the addition of a dispersant will have the consequence of promoting the solubilization of the lightest hydrocarbon molecules and increasing the bioavailability of the petroleum with respect to the attack of the microorganisms, and thus to accelerate its biodegradation. However, this theoretical approach results from temperate and surface experiences. The background on Arctic regions is limited in comparison to temperate areas and is very marginal in deep-sea areas.

5.5.1. *Prominence of oil biodegradation for ecosystem recovery*

The total recovery of polluted ecosystems is possible only through a biodegradation phenomenon. Oil biodegradation is a biological process where harmful hydrocarbon compounds are metabolized in non-hazardous molecules before being used as a carbon source and remineralized [MAR 01]. Oil biodegradation is a natural phenomenon generally mediated by autochthonous microbiota [HEA 06]. Oil is mainly biodegraded by prokaryote and, to a lesser extent, by fungi and microalgae [HEA 06, MCG 12]. Oil biodegradation occurs in various natural or artificial environments as long as there is an oil/water interface [HEA 03]. To date, oil is shown to be systematically biodegraded in any environment studied whose temperature ranges between –1 and 85°C, from the surface oil slick to the alpine lake through an artificial ecosystem like a fuel tank [MAR 01, AIT 04, MCF 14]. Various catabolic pathways have been demonstrated [ATL 81, WID 01, AIT 04], and it seems that there is no general rule with enzymes involved in oil biodegradation phenomenon. A probable reason is the very ancient emergence of such a metabolism [HEA 06] whose effect is dramatically increased by horizontal gene transfers between microorganisms [VAN 03]. Microorganisms have thus evolved several strategies to degrade hydrocarbon including the production of secondary metabolites such as biosurfactants [RON 02]. Biodegradation begins with the small to medium chemical species (C_1–C_{35} alkanes, small aromatics) which are more metabolically available for the microbiota as a carbon source [HEA 06, ROJ 09]. When the pool of small hydrocarbon species is depleted, the biodegradation naturally shifts to branched alkanes and larger as well as alkylated aromatics [HEA 06]. The consumption of small hydrocarbon species can lead to an increase in the viscosity of crude oil [AIT 04] or the quality loss of refined products such as biofuel [SCH 09]. If its chemical composition can be considered as the main factor affecting the oil fate, oil biodegradation is also strongly related to the abiotic factors of the environment.

5.5.2. *Abiotic factors influencing oil biodegradation*

– *Temperature* can influence the biodegradation process through its effect on the viscosity, diffusion, volatility and solubility of oil, and thus on the bioavailability of hydrocarbon species [MAR 01]. So far, except for oil reservoirs that were paleopasteurized during burial [WIL 01], it is considered that, in nature, biodegradation systematically occurs between –1 and 85°C [MCF 14, AIT 04, HOL 11]. With more than 55 publications since 1975, temperature has probably been the most studied abiotic factor influencing oil biodegradation. Microorganisms have an oil-degrading enzymatic apparatus whose optimum functioning is temperature-dependent. Psychrophilic microorganisms are responsible for biodegradation at a low temperature (–1 to 20°C), mesophilic microorganisms degrade hydrocarbon between 15 and 40°C, thermophilic microorganisms between 40 and 80°C and hyperthermophilic microorganisms beyond 80°C [MAR 01].

– *Sunlight-mediated photooxidation* positively affects the biodegradation phenomenon by converting aromatic compounds such as PAH in polar species, thus increasing their bioavailability for biodegraders [PRI 03]. At the same time, it can generate toxic compounds for non-hydrocarbonoclast organisms [KIN 14]. Photooxidation has been reported at all latitudes, including the polar environment [BRA 08]. It can allow 36% of the total oil fraction to be removed from seawater [DUT 00].

– *Environmental pH* seems to have a limited effect on oil biodegradation. Acidophilic microorganisms can degrade hydrocarbon at a low pH value down to pH 2 [RÖL 10] and alkaliphilic microorganisms at a high pH value up to pH 10 [FAH 08, SOR 12]. Nevertheless, in the literature, a large part of the oil-degrading microorganism has an optimal pH growth near pH 8, most likely because a lot of biological models are marine microorganisms [HEA 06].

– *Oxygenation* is known to accelerate biodegradation, especially in aquatic environments [ATL 81, LEA 90]. However, anaerobic conditions are not detrimental to oil biodegradation [AIT 04]. Many studies have reported the anaerobic biodegradation of a vast variety of hydrocarbon compounds by prokaryotes [COA 97, GRI 00, SPO 00]. Aerobic degradation allows a higher metabolic rate than anaerobic degradation [ATL 91] but, for example, in an aquatic environment, oxygenation can also be linked to the synergistic effect of a better diffusion of hydrocarbon through water movement which enhances bioavailability in return [SAN 00, LEE 13].

– *Osmolarity*, sometimes called ionic strength, does not drastically affect biodegradation. Many studies report the existence of hallophilic prokaryotes that are able to digest oil and derivatives [MAR 01, LEB 08]. Bacteria are better represented than Archaea and often exhibit other extremophilic life traits such as thermophily or hyperthermophily [MCG 10, FAT 14].

– *Nutrient availability*, particularly regarding nitrogen (N), phosphorus (P) and iron (Fe), has revealed to be crucial for the biodegradation of oil [LEA 90]. This phenomenon is commonly used as an asset in the bioremediation processes (biostimulation) of an oil polluted environment with xenobiotic-degrading bacteria or by bioaugmentation using autochthonous bacteria [PRI 97, WRI 04]. Indeed, fertilization greatly improves the recovery of the environment [COU 07, TYA 11]. In return, N and P depleted environments such as tropical waters or Fe depleted regions like the southern ocean can be more affected by oil spills [ATL 72, DIB 76, DEL 02].

– *Environmental redox potential* is generally strongly related to oxygenation and pH in the environment. Redox potential can impact the chemical speciation of trace elements and Fe, particularly in sediments [DEL 09], and can limit the biomass of biodegraders or drive to less efficient catabolic pathways [DEL 80, HAM 80].

– *Hydrostatic pressure* is probably the least studied abiotic factor influencing oil biodegradation [SME 17]. Paradoxically, 75% of the Earth's area potentially subject to oil pollution is situated below 500 m of water depth [CAN 14], and oil fields are naturally subject to high pressure [KOT 12]. Several oil-degrading piezotolerant microorganisms (viable at a pressure of > 50 MPa) or piezophilic microorganisms (with an optimum growth at a pressure of > 50 MPa) from sub-seafloor or oil reservoirs have been already isolated [TAP 10, SCH 17]. However, recent experimental studies pointed out the deleterious effect of hydrostatic pressure on the oil biodegradation ability of piezophilic bacteria [SCH 14] as well as on the structuration of microbial community [FAS 18]. The lack of data about the effect of hydrostatic pressure on oil biodegradation underlines the need for greater research efforts on this topic.

5.6. Perspectives in a context of oil spill in polar and deep waters

The rapid depletion of historical oil wells coupled with a constant increase in oil costs and demand drives oil companies to consider the prospects of new oil fields. These next-generation oil fields are often located to inhospitable places such as the polar environment (e.g. Prirazlomnoe, belonging to Gazprom) or the deep-sea floor (e.g. *Deepwater Horizon*, exploited by BP) or both (e.g. Burger J, property of Shell). To make the extraction of oil lucrative, these new site exploitations require technological breakthroughs where the current low maturity can lead to industrial accident. Oil spills in polar or deep environments, and particularly in polar deep waters, can be quite devastating and very problematic to remediate [ATL 11]. In a context of global change (GC) and subsequent consequences (e.g. the opening of the North Eastern Road in the Arctic), or the generalization of ultra-deep water oil wells (–1,000 m and deeper) like the *Deepwater Horizon* drilling rig, the probabilities of huge accidental oil spills in these extremely sensitive ecological areas could dramatically increase.

This ongoing problem suffers from a big gap of knowledge:

– First, the technical responses in these habitats are very difficult to set on. If the use of dispersant in the marine environment remains an effective technique to accelerate the dilution of oil and limit massive groundings of hydrocarbons on the coasts, many questions remain unresolved concerning its use in extreme environments. The influence of temperature, salinity, hydrostatic pressure and their synergistic effects on dispersants is still under-documented. The *Deepwater Horizon* oil spill was the perfect example of approximations in this domain (see section 5.3). In order to limit the amount of oil at the sea surface, dispersants were used in the deep sea without any consideration for the biocenosis of the deep sea at both benthic and pelagic zones, not only in the short term but also in the long term. Indeed, oil biodegradation in the lowest layers of the ocean was under-documented.

– To date, the vast majority of studies about biodegradation have focused on a specific ecophysiological character of microbial biodegraders. To illustrate this, from an oil biodegradation point of view, polar and deep waters share a common characteristic: temperatures below 4°C. But considering the temperature as the only abiotic factor influencing biodegradation would be reducing and tricky. So far, no extensive studies have explored the synergistic/antagonistic effects of *all* the abiotic factors on biodegradation phenomenon in this kind of environment. Except photooxidation, microbial biodegraders from polar or deep waters are subject to the same set of abiotic factors but with different intensities/variations. An accidental oil spill would most probably impact the sea ice microbial community (SIMCO) quite differently than the deep-sea floor microbiota.

– Biodegraders of the SIMCO are subject to seasonal variations through the sea ice *embâcle/debâcle*[1] cycle. These variations affect the photooxidation, the osmolarity through the formation of brines and sea ice, the oxygenation and, to a lesser extent, the temperature [GER 05]. Oil biodegradation is directly affected by this seasonal cycle [GAR 03, BRA 08]. In comparison, deep-sea-floor microbial degraders have a very homogeneous environment with no photooxidation but the hydrostatic pressure is a prominent abiotic factor.

1 Jam/breakdown.

Finally, these two communities would be subject to different kinds of oil pollution in function of the depth of a hypothetical accident [ATL 11].

A better comprehension of the biodegradation mechanisms at play and the relationships between the different microbial compartments of polar and deep waters is necessary. To accurately estimate ecosystem recovery capabilities, future researches should imperatively take into account all the set of abiotic factors (e.g. hydrostatic pressure) affecting the oil-biodegraders microbial communities and focus on their possible synergistic/antagonistic effects.

The new oil research areas are necessary to provide an efficient supply to anthropic needs but raise a lot of concern. Indeed, an increase in human activities in these fragile ecosystems could cause irreversible damages such as loss of biodiversity and even destruction of some isolated ecosystem. The Arctic ecosystem is more studied than the deep-sea ecosystem, and this survey demonstrates the strong impact of human activities. Several examples due to global change have already deeply disturbed the Arctic area such as an increase in subarctic species presence [HIG 09, PRO 12] and the apparition of new parasites or diseases [DAV 11].

A better response would be the reduction of society's needs for petroleum products. However, in order to reduce the risk of impact in these environments, it is necessary to continue the efforts already undertaken in both the field of "clean" oil production and in the anti-pollution control protocols. Finally, an increase in research studies in these areas will allow a better understanding of their functioning and consequently will help to point out better technical response strategies for future and, unfortunately, predictable oil pollution.

5.7. Bibliography

[ABE 01] ABE F., HORIKOSHI K., "The biotechnological potential of piezophiles", *Trends in Biotechnology*, vol. 19, no. 3, pp. 102–108, 2001.

[AIT 04] AITKEN C.M., JONES D.M., LARTER S.R., "Anaerobic hydrocarbon biodegradation in deep subsurface oil reservoirs", *Nature*, vol. 431, no. 7006, pp. 291–294, 2004.

[AND 06] ANDERSON T.R., RICE T., "Deserts on the sea floor: Edward Forbes and his azoic hypothesis for a lifeless deep ocean", *Endeavour*, vol. 30, no. 4, pp. 131–137, 2006.

[ATL 72] ATLAS R.M., BARTHA R., "Degradation and mineralization of petroleum in sea water: limitation by nitrogen and phosphorous", *Biotechnology and Bioengineering*, vol. 14, no. 3, pp. 309–318, 1972.

[ATL 81] ATLAS R.M., "Microbial degradation of petroleum hydrocarbons: an environmental perspective", *Microbiological Reviews*, vol. 45, no. 1, p. 180, 1981.

[ATL 91] ATLAS R.M., "Microbial hydrocarbon degradation-biorem ediation of oil spills", *Journal of Chemical Technology and Biotechnology*, vol. 52, no. 2, pp. 149–156, 1991.

[ATL 11] ATLAS R.M., HAZEN T.C., "Oil biodegradation and bioremediation: a tale of the two worst spills in US history", *Environmental Science and Technology*, vol. 45, no. 16, pp. 6709–6715, doi: 10.1021/es2013227, 2011.

[ATS 95] ATSDR, *Toxicological Profile for Polycyclic Aromatic Hydrocarbons*, U.S. Department of Health and Human Services, Atlanta, GA, 1995.

[BEJ 17] BEJARANO A.C., GARDINER W.W., BARRON M.G. *et al.*, "Relative sensitivity of Arctic species to physically and chemically dispersed oil determined from three hydrocarbon measures of aquatic toxicity", *Marine Pollution Bulletin*, vol. 122, nos 1–2, pp. 316–322, 2017.

[BRA 08] BRAKSTAD O.G., NONSTAD I., FAKSNESS L.G. *et al.*, "Responses of microbial communities in Arctic sea ice after contamination by crude petroleum oil", *Microbial Ecology*, vol. 55, no. 3, pp. 540–552, 2008.

[BRO 11] BROWN A., THATJE S., "Respiratory response of the deep-sea amphipod *Stephonyx biscayensis* indicates bathymetric range limitation by temperature and hydrostatic pressure", *PLoS One*, vol. 6, no. 12, p. e28562, doi.org/10.1371/journal.pone.0028562, 2011.

[BRO 14a] BROJE V., GALA W., NEDWED T. *et al.*, "Consensus on the state of the knowledge and research recommendations on the fate and effects of deep water releases of oil, dispersants and dispersed oil", *International Oil Spill Conference Proceedings May 2014*, vol. 2014, no. 1, pp. 225–237, doi: 10.7901/2169-3358-2014.1.225, 2014.

[BRO 14b] BROWN A., THATJE S., "Explaining bathymetric diversity patterns in marine benthic invertebrates and demersal fishes: physiological contributions to adaptation of life at depth", *Biological Reviews*, vol. 89, no. 2, pp. 406–426, doi: 10.1111/brv.12061, 2014.

[CAM 05] CAMUS L., GULLIKSEN B., "Antioxidant defense properties of Arctic amphipods: comparison between deep-, sublittoral and surface-water species", *Marine Biology*, vol. 146, no. 2, pp. 355–362, 2005.

[CAM 10] CAMILLI R., REDDY C.M., YOERGER D.R. *et al.*, "Tracking hydrocarbon plume transport and biodegradation at Deepwater Horizon", *Science*, vol. 330, no. 6001, pp. 201–204, 2010.

[CAN 14] CANGANELLA F., KATO C., *Deep-Ocean Ecosystems*, John Wiley & Sons Ltd, Chichester, available at: http://www.els.net, doi:10.1002/9780470015902.a0003192.pub2, January 2014.

[CHA 05] CHAPMAN P.M., RIDDLE M.J., "Toxic effects of contaminants in polar marine environments", *Environment Science Technology*, vol. 5, pp. 200–207, 2005.

[CHA 07] CHAPMAN H., PURNELL K., LAW R.J. *et al.*, "The use of chemical dispersants to combat oil spills at sea: a review of practice and research needs in Europe", *Marine Pollution Bulletin*, vol. 54, pp. 827–838, 2007.

[CHR 95] CHRISTIANSEN J.S., GEORGE S., "Contamination of food by crude oil affects food selection and growth performance, but not appetite, in an Arctic fish, the polar cod (Boreogadus saida)", *Polar Biology*, vol. 15, pp. 277–281, doi: 10.1007/BF00239848, 1995.

[CHR 10] CHRISTIANSEN J.S., KARAMUSHKO L.I., NAHRGANG J., "Sub-lethal levels of waterborne petroleum may depress routine metabolism in polar cod Boreogadus saida (Lepechin, 1774)", *Polar Biology*, vol. 33, no. 8, pp. 1049–1055, 2010.

[CLA 79] CLARKE A., "On living in cold water: K-strategies in Antarctic benthos", *Marine Biology*, vol. 55, pp. 111–119, 1979.

[CLA 13] CLAIREAUX G., THERON M., PRINEAU M. *et al.*, "Effects of oil exposure and dispersant use upon environmental adaptation performance and fitness in the European sea bass, Dicentrarchus labrax", *Aquatic Toxicology*, vols 130–131, pp. 160–170, 2013.

[COA 97] COATES J.D., WOODWARD J., ALLEN J. *et al.*, "Anaerobic degradation of polycyclic aromatic hydrocarbons and alkanes in petroleum-contaminated marine harbor sediments", *Applied and Environmental Microbiology*, vol. 63, no. 9, pp. 3589–3593, 1997.

[COM 08] COMISO J.C., PARKINSON C.L., GERSTEN R. *et al.*, "Accelerated decline in the Arctic sea ice cover", *Geophysical Research Letters*, vol. 35, no. 1, pp. 1–6, doi: 10.1029/2007GL031972, 2008.

[COT 12] COTTIN D., BROWN A., OLIPHANT A. *et al.*, "Sustained hydrostatic pressure tolerance of the shallow water shrimp Palaemonetes varians at different temperatures: insights into the colonisation of the deep sea", *Comparative Biochemistry and Physiology Part A: Molecular and Integrative Physiology*, vol. 162, pp. 357–363, 2012.

[COU 07] COULON F., MCKEW B.A., OSBORN A.M. *et al.*, "Effects of temperature and biostimulation on oil-degrading microbial communities in temperate estuarine waters", *Environmental Microbiology*, vol. 9, no. 1, pp. 177–186, 2007.

[DAV 92] DAVISON W., FRANKLIN C.E., MCKENZIE J.C. *et al.*, "The effects of acute exposure to the water soluble fraction of diesel fuel oil on survival and metabolic rate of an antarctic fish (Pagothenia borchgrevinki)", *Comparative Biochemistry and Physiology Part C*, vol. 102, no. 1, pp. 185–188, 1992.

[DAV 11] DAVIDSON R., SIMARD M., KUTZ S. *et al.*, "Arctic parasitology: why should we care?", *Trends in Parasitology*, vol. 27, no. 6, pp. 239–245, 2011.

[DEB 77] DE BROYER C., "Analysis of gigantism and dwarfness of antarctic and sub-antarctic gammaridean amphipoda", in LLANO G. (ed.), *Adaptations within Antarctic Ecosystems*, Smithsonian Institution Press, Washington, DC, 1977.

[DEL 80] DELAUNE R.D., HAMBRICK III G.A., PATRICK JR W.H., "Degradation of hydrocarbons in oxidized and reduced sediments", *Marine Pollution Bulletin*, vol. 11, no. 4, pp. 103–106, 1980.

[DEL 02] DELILLE D., DELILLE B., PELLETIER E., "Effectiveness of bioremediation of crude oil contaminated subantarctic intertidal sediment: the microbial response", *Microbial Ecology*, vol. 44, no. 2, pp. 118–126, 2002.

[DEL 09] DELL'ANNO A., BEOLCHINI F., GABELLINI M. *et al.*, "Bioremediation of petroleum hydrocarbons in anoxic marine sediments: consequences on the speciation of heavy metals", *Marine Pollution Bulletin*, vol. 58, no. 12, pp. 1808–1814, 2009.

[DEH 11] DE HOOP L., SCHIPPER A., LEUVEN R. *et al.*, "Sensitivity of polar and temperate marine organisms to oil components", *Environmental Science and Technology*, vol. 45, no. 20, pp. 9017–9023, 2011.

[DES 79] DE SMEDT H., BORGHGRAEF R., CEUTERICK F. *et al.*, "Pressure effects on lipid-protein interactions in ($Na^+ + K^+$)-ATPase", *Biochimica and Biophysica Acta*, vol. 556, pp. 479–489, 1979.

[DIB 76] DIBBLE J.T., BARTHA R., "Effect of iron on the biodegradation of petroleum in seawater", *Applied and Environmental Microbiology*, vol. 31, no. 4, pp. 544–550, 1976.

[DOD 07] DODDS L., ROBERTS J., TAYLOR A. *et al.*, "Metabolic tolerance of the cold-water coral Lophelia pertusa (Scleractinia) to temperature and dissolved oxygen change", *Journal of Experimental Marine Biology and Ecology*, vol. 349, no. 2, pp. 205–214, 2007.

[DUS 14a] DUSSAUZE M., CAMUS L., LE FLOCH S. *et al.*, "Impact of dispersed fuel oil on cardiac mitochondrial function on polar cod Boreogadus saida", *Environmental Science and Pollution Research*, vol. 21, no. 24, pp. 13779–13788, 2014.

[DUS 14b] DUSSAUZE M., CAMUS L., LE FLOCH S. *et al.*, "Effect of dispersed oil on fish tissue respiration, a comparison between a temperate (Dicentrarchus labrax) and an Arctic (Boreogadus saida) species", *Proceedings of the 37th AMOP Technical Seminar on Environmental Contamination and Response*, Environment and Climate Change Canada, pp. 482–493, Ottawa, 2014.

[DUS 15a] DUSSAUZE M., PICHAVANT-RAFINI K., LE FLOCH S. *et al.*, "Acute toxicity of chemically and mechanically dispersed crude oil on juvenile sea bass (Dicentrarchus labrax): absence of synergistic effects between oil and dispersants", *Environmental Toxicology and Chemistry*, vol. 34, no. 7, pp. 1543–1551, 2015.

[DUS 15b] DUSSAUZE M., DANION M., LE FLOCH S. *et al.*, "Growth and immune system performance to assess the impact of dispersed oil on sea bass (Dicentrarchus labrax)", *Ecotoxicological and Environmental Safety*, vol. 120, pp. 215–222, 2015.

[DUS 16a] DUSSAUZE M., PICHAVANT-RAFINI K., LE FLOCH S. *et al.*, "In vitro diving simulation: a new approach to assess biological impact of hydrocarbons at depth", *Proceedings of the 39th AMOP Technical Seminar*, Environment and Climate Change Canada, Ottawa, pp. 304–316, 2016.

[DUS 16b] DUSSAUZE M., PICHAVANT-RAFINI K., BELHOMME M. *et al.*, "Dispersed oil decreases the ability of a model fish (Dicentrarchus labrax) to cope with hydrostatic pressure", *Environmental Science and Pollution Research*, vol. 24, no. 3, pp. 3054–3062, doi: 10.1007/s11356-016-7955-8, 2016.

[DUT 00] DUTTA T.K., HARAYAMA S., "Fate of crude oil by the combination of photooxidation and biodegradation", *Environmental Science and Technology*, vol. 34, no. 8, pp. 1500–1505, 2000.

[FAH 08] FAHY A., BALL A.S., LETHBRIDGE G. *et al.*, "Isolation of alkali-tolerant benzene-degrading bacteria from a contaminated aquifer", *Letters in Applied Microbiology*, vol. 47, no. 1, pp. 60–66, 2008.

[FAS 18] FASCA H., CASTILHO L.V., CASTILHO J.F.M. *et al.*, "Response of marine bacteria to oil contamination and to high pressure and low temperature deep sea conditions", *MicrobiologyOpen*, vol. 7, pp. 1–10. doi.org/10.1002/mbo3.550, 2018.

[FAT 14] FATHEPURE B.Z., "Recent studies in microbial degradation of petroleum hydrocarbons in hypersaline environments", *Frontiers in Microbiology*, vol. 5, p. 173, 2014.

[FEL 03] FELLER G., GERDAY C., "Psychrophilic enzymes: hot topics in cold adaptation", *Nature Reviews Microbiology*, vol. 1, no. 3, pp. 200–208, 2003.

[GAG 92] GAGE J.D., TYLER P.A., *Deep-sea Biology: A Natural History of Organisms at the Deep-sea Floor*, Cambridge University Press, Cambridge, 1992.

[GAR 03] GARRETT R.M., ROTHENBURGER S.J., PRINCE R.C., "Biodegradation of fuel oil under laboratory and arctic marine conditions", *Spill Science & Technology Bulletin*, vol. 8, no. 3, pp. 297–302, 2003.

[GAR 13] GARDINER W.W., WORD J.Q., WORD J.D. *et al.*, "The acute toxicity of chemically and physically dispersed crude oil to key arctic species under arctic conditions during the open water season", *Environmental Toxicology and Chemistry*, vol. 32, no. 10, pp. 2284–2300, 2013.

[GER 05] GERDES B., BRINKMEYER R., DIECKMANN G. *et al.*, "Influence of crude oil on changes of bacterial communities in Arctic sea-ice", *FEMS Microbiology Ecology*, vol. 53, no. 1, pp. 129–139, 2005.

[GIA 01] GIANESE G., ARGOS P., PASCARELLA S., "Structural adaptation of enzymes to low temperatures", *Protein Engineering*, vol. 14, no. 3, pp. 141–148, 2001.

[GOM 04] GOMES J., STEINER W., "The biocatalytic potential of extremophiles and extremozymes", *Food Technology and Biotechnology*, vol. 42, no. 4, pp. 223–235, 2004.

[GRI 00] GRISHCHENKOV V.G., TOWNSEND R.T., MCDONALD T.J. *et al.*, "Degradation of petroleum hydrocarbons by facultative anaerobic bacteria under aerobic and anaerobic conditions", *Process Biochemistry*, vol. 35, no. 9, pp. 889–896, 2000.

[GRI 12] GRIFFITHS S.K., "Oil release from Macondo Well MC252 following the Deepwater Horizon Accident", *Environmental Science and Technology*, vol. 46, no. 10, pp. 5616–5622, 2012.

[HAM 80] HAMBRICK G.A., DELAUNE R.D., PATRICK W.H., "Effect of estuarine sediment pH and oxidation-reduction potential on microbial hydrocarbon degradation", *Applied and Environmental Microbiology*, vol. 40, no. 2, pp. 365–369, 1980.

[HAN 12] HANSEN B.H., ALTIN D., OLSEN A.J. *et al.*, "Acute toxicity of naturally and chemically dispersed oil on the filter-feeding copepod Calanus finmarchicus", *Ecotoxicology and Environmental Safety*, vol. 86, pp. 38–46, 2012.

[HAN 13] HANSEN B.H., ALTIN D., ØVERJORDET I.B. *et al.*, "Acute exposure of water soluble fractions of marine diesel on Arctic Calanus glacialis and boreal Calanus finmarchicus: effects on survival and biomarker response", *Science of the Total Environment*, vol. 449, no. 1, pp. 276–284, 2013.

[HAN 14] HANSEN B.H., ALTIN D., BONAUNET K. *et al.*, "Acute toxicity of eight oil spill response chemicals to temperate, boreal, and arctic species", *Journal of Toxicology and Environmental Health Part A*, vol. 77, nos 9–11, pp. 495–505, 2014.

[HAZ 95] HAZEL J.R., "Thermal adaptation in biological membranes: is homeoviscous adaptation the explanation?", *Annual Review of Physiology*, vol. 57, no. 1, pp. 19–42, 1995.

[HAZ 10] HAZEN T.C., DUBINSKY E.A., DESANTIS T.Z. *et al.*, "Deep-sea oil plume enriches indigenous oil-degrading bacteria", *Science*, vol. 330, no. 6001, pp. 204–208, 2010.

[HEA 03] HEAD I.M., JONES D.M., LARTER S.R., "Biological activity in the deep subsurface and the origin of heavy oil", *Nature*, vol. 426, no. 6964, p. 344, 2003.

[HEA 06] HEAD I.M., JONES D.M., RÖLING W.F., "Marine microorganisms make a meal of oil", *Nature Reviews Microbiology*, vol. 4, no. 3, pp. 173–182, 2006.

[HIG 09] HIGDON J.W., FERGUSON S.H., "Loss of Arctic sea ice causing punctuated change in sightings of killer whales (Orcinus orca) over the past century", *Ecological Applications*, vol. 19, no. 5, pp. 1365–1375, 2009.

[HIR 96] HIRCHE H., "Diapause in the marine copepod, Calanus finmarchicus: a review", *Ophelia*, vol. 44, nos 1–3, pp. 129–143, 1996.

[HJO 11] HJORTH M., NIELSEN T.G., "Oil exposure in a warmer Arctic: potential impacts on key zooplankton species", *Marine Biology*, vol. 158, no. 6, pp. 1339–1347, 2011.

[HOL 11] HOLMES D.E., RISSO C., SMITH J.A. *et al.*, "Anaerobic oxidation of benzene by the hyperthermophilic archaeon Ferroglobus placidus", *Applied and Environmental Microbiology*, vol. 77, no. 17, pp. 5926–5933, 2011.

[HOP 13] HOP H., GJØSÆTER H., "Polar cod (Boreogadus saida) and capelin (Mallotus villosus) as key species in marine food webs of the Arctic and the Barents Sea", *Marine Biology Research*, vol. 9, no. 9, pp. 878–894, 2013.

[IEA 13] IEA, *World Energy Outlook 2013*, OECD/IEA, Paris, 2013.

[JON 10] JONSSON H., SUNDT R.C., AAS E. *et al.*, "The arctic is no longer put on ice: evaluation of polar cod (Boreogadus saida) as a monitoring species of oil pollution in cold waters", *Marine Pollution Bulletin*, vol. 60, no. 3, pp. 390–395, 2010.

[JOY 15] JOYE S.B., "Deepwater Horizon, 5 years on", *Science*, vol. 349, no. 6248, pp. 592–593, 2015.

[KIN 14] KING S.M., LEAF P.A., OLSON A.C. *et al.*, "Photolytic and photocatalytic degradation of surface oil from the Deepwater Horizon spill", *Chemosphere*, vol. 95, pp. 415–422, 2014.

[KOT 12] KOTLAR H.K., "Extreme to the fourth power! Oil-, high temperature-, salt- and pressure-tolerant microorganisms in oil reservoirs. What secrets can they reveal?", in ANITORI R.P. (ed.), *Extremophiles: Microbiology and Biotechnology*, Caister Academic Press, Norfolk, 2012.

[KOV 11] KOVACS K.M., LYDERSEN C., OVERLAND J.E. *et al.*, "Impacts of changing sea-ice conditions on Arctic marine mammals", *Marine Biodiversity*, vol. 41, no. 1, pp. 181–194, doi: 10.1007/s12526-010-0061-0, 2011.

[KOY 04] KOYAMA J., KAKUNO A., "Toxicity of heavy fuel oil, dispersant, and oil-dispersant mixtures to a marine fish, Pagrus major", *Fisheries Science*, vol. 70, no. 4, pp. 587–594, 2004.

[KUJ 11] KUJAWINSKI E.B., KIDO SOULE M.C., VALENTINE D.L. *et al.*, "Fate of dispersants associated with the Deepwater Horizon oil spill", *Environmental Science and Technology*, vol. 45, no. 4, pp. 1298–1306, 2011.

[LAS 90] LASSITER R.R., HALLAM T.G., "Survival of the fattest: implications for acute effects of lipophilic chemicals on aquatic populations", *Environmental Toxicology and Chemistry*, vol. 9, no. 5, pp. 585–595, 1990.

[LEA 90] LEAHY J.G., COLWELL R.R., "Microbial degradation of hydrocarbons in the environment", *Microbiological Reviews*, vol. 54, no. 3, pp. 305–315, 1990.

[LEB 08] LE BORGNE S., PANIAGUA D., VAZQUEZ-DUHALT R., "Biodegradation of organic pollutants by halophilic bacteria and archaea", *Journal of Molecular Microbiology and Biotechnology*, vol. 15, nos 2–3, pp. 74–92, 2008.

[LEE 06] LEE R.F., HAGEN W., KATTNER G., "Lipid storage in marine zooplankton", *Marine Ecological Progress Series*, vol. 307, pp. 273–306, 2006.

[LEE 13] LEE K., NEDWED T., PRINCE R.C. *et al.*, "Lab tests on the biodegradation of chemically dispersed oil should consider the rapid dilution that occurs at sea", *Marine Pollution Bulletin*, vol. 73, no. 1, pp. 314–318, 2013.

[LEF 99] LE FLOCH S., MERLIN F., GUILLERME M. *et al.*, "A field experimentation on bioremediation: BIOREN", *Environmental Technology*, vol. 20, no. 8, pp. 897–907, 1999.

[LEF 14] LE FLOCH S., DUSSAUZE M., MERLIN F.X. *et al.*, "DISCOBIOL: assessment of the impact of dispersant use for oil spill response in coastal or estuarine areas", *International Oil Spill Conference Proceedings of May 2014*, vol. 2014, no. 1, pp. 491–503, doi: http://dx.doi.org/10.7901/2169-3358-2014.1.491, 2014.

[LES 00] LESSARD R.R., DEMARCO G., "The significance of oil spill dispersants", *Spill Science & Technology Bulletin*, vol. 6, no. 1, pp. 59–68, 2000.

[MAC 84] MACDONALD A.G., "The effects of pressure on the molecular structure and physiological functions of cell membrane", *Philosophical Transactions of the Royal Society of London, Series B*, vol. 304, no. 1118, pp. 47–68, 1984.

[MAC 97] MACDONALD A.G., "Hydrostatic pressure as an environmental factor in life processes", *Comparative Biochemistry and Physiology Part A: Molecular and Integrative Physiology*, vol. 116, no. 4, pp. 291–297, 1997.

[MAC 12] MACKEY A.P., ATKINSON A., HILL S.L. *et al.*, "Antarctic macrozooplankton of the southwest Atlantic sector and Bellingshausen Sea: baseline historical distributions (discovery investigations, 1928–1935) related to temperature and food, with projections for subsequent ocean warming", *Deep-Sea Research Part II: Topical Studies in Oceanography*, vols 59–60, pp. 130–146, 2012.

[MAR 01] MARGESIN R., SCHINNER F., "Biodegradation and bioremediation of hydrocarbons in extreme environments", *Applied Microbiology and Biotechnology*, vol. 56, no. 5, pp. 650–663, 2001.

[MCC 15] MCCLAIN C.R., SCHLACHER T.A., "On some hypotheses of diversity of animal life at great depths on the sea floor", *Marine Ecology*, vol. 36, no. 4, pp. 849–872, 2015.

[MCF 14] MCFARLIN K.M., PRINCE R.C., PERKINS R. *et al.*, "Biodegradation of dispersed oil in arctic seawater at – 1 C". *PloS One*, vol. 9, no. 1, p. e84297, 2014.

[MCG 10] MCGENITY T.J., "Halophilic hydrocarbon degraders", in *Handbook of Hydrocarbon and Lipid Microbiology*, Springer, Berlin, 2010.

[MCG 12] MCGENITY T.J., FOLWELL B.D., MCKEW B.A. *et al.*, "Marine crude-oil biodegradation: a central role for interspecies interactions", *Aquatic Biosystems*, vol. 8, no. 1, p. 10, 2012.

[MES 13] MESTRE N.C., CALADO R., SOARES A., "Exploitation of deep-sea resources: the urgent need to understand the role of high pressure in the toxicity of chemical pollutants to deep-sea organisms", *Environmental Pollution*, vol. 185, pp. 369–371, doi: 10.1016/j.envpol.2013.10.02, 2013.

[MIL 11] MILINKOVITCH T., KANAN R., THOMAS-GUYON H. *et al.*, "Effects of dispersed oil exposure on the bioaccumulation of polycyclic aromatic hydrocarbons and mortality of juvenile Liza ramada", *Science of the Total Environment*, vol. 409, no. 9, pp. 1643–1650, 2011.

[MON 13] MONTAGNA P.A., BAGULEY J.G., COOKSEY C. *et al.*, "Deep-sea benthic footprint of the Deepwater Horizon blowout", *PLoS One*, vol. 8, no. 8, p. e70540, doi: 10.1371/journal.pone.0070540, 2013.

[MUE 13] MUEHLENBACHS L., COHEN M., GERARDEN T., "The impact of water depth on safety and environmental performance in offshore oil and gas production", *Energy Policy A*, vol. 55, pp. 699–705, 2013.

[NAH 10a] NAHRGANG J., CAMUS L., BROMS F. *et al.*, "Seasonal baseline levels of physiological and biochemical parameters in polar cod (Boreogadus saida): implications for environmental monitoring", *Marine Pollution Bulletin*, vol. 60, no. 8, pp. 1336–1345, 2010.

[NAH 10b] NAHRGANG J., CAMUS L., CARLS M.G. *et al.*, "Biomarker responses in polar cod (Boreogadus saida) exposed to the water soluble fraction of crude oil", *Aquatic Toxicology*, vol. 97, no. 3, pp. 234–242, 2010.

[NOR 12] NORSE E.A. *et al.*, "Sustainability of deep-sea fisheries", *Marine Policy*, vol. 36, no. 2, pp. 307–320, 2012.

[OHM 13] OHMAE E., MIYASHITA Y., KATO C., "Thermodynamic and functional characteristics of deep-sea enzymes revealed by pressure effects", *Extremophiles*, vol. 17, no. 5, pp. 701–709, 2013.

[OLS 11] OLSEN G.H., SMIT M.G., CARROLL J. *et al.*, "Arctic versus temperate comparison of risk assessment metrics for 2-methyl-naphthalene", *Marine Environmental Research*, vol. 72, no. 4, pp. 179–187, 2011.

[OLS 16] OLSEN G.H., COQUILLÉ N., LE FLOCH S. *et al.*, "Sensitivity of the deep-sea amphipod Eurythenes gryllus to chemically dispersed oil", *Environmental Science and Pollution Research*, vol. 23, no. 7, pp. 6497–6505, 2016.

[OLU 04] OLU-LE ROY K., SIBUETA M., FIALA-MÉDIONI A. *et al.*, "Cold seep communities in the deep eastern Mediterranean Sea: composition, symbiosis and spatial distribution on mud volcanoes", *Deep Sea Research Part I: Oceanographic Research Papers*, vol. 51, no. 12, pp. 1915–1936, 2004.

[ORL 09] ORLOVA E.L., DOLGOV A.V., RUDNEVA G.B. *et al.*, "Trophic relations of capelin Mallotus villosus and polar cod Boreogadus saida in the Barents Sea as a factor of impact on the ecosystem", *Deep Sea Research Part II: Topical Studies in Oceanography*, vol. 56, nos 21–22, pp. 2054–2067, 2009.

[OVE 14] OVERLAND J., WANG M., WALSH J. *et al.*, "Future Arctic climate changes: adaptation and mitigation time scales", *Earth's Future*, vol. 2, no. 2, pp. 68–74, 2014.

[PAR 03] PARMESAN C., YOHE G., "A globally coherent fingerprint of climate change impacts across natural systems", *Nature*, vol. 421, no. 6918, pp. 37–42, 2003.

[PEC 02] PECK L.S., "Ecophysiology of antarctic marine ectotherms: limits to life", *Polar Biology*, vol. 25, no. 1, pp. 31–40, 2002.

[PER 08] PEROVICH D.K., RITCHER-MENGE J.A., JONES K.F. *et al.*, "Sunlight, water, and ice: extreme arctic sea ice melt during the summer of 2007", *Geophysical Research Letters*, vol. 35, no. 11, pp. 1–4, doi: 10.1029/2008GL034007, 2008.

[PIE 08] PIETRI D., SOULE A., KERSHNER J. *et al.*, "The Arctic shipping and environmental management agreement: a regime for marine pollution", *Coastal Management*, vol. 36, no. 5, pp. 508–523, 2008.

[POS 13] POST E., BHATT U.S., BITZ C.M. *et al.*, "Ecological consequences of sea-ice decline", *Science*, vol. 341, no. 6145, pp. 519–524, 2013.

[PRA 07] PRADILLON F., GAILL F., "Pressure and life: some biological strategies", *Reviews in Environmental Science and Bio/Technology*, vol. 6, nos 1–3, pp. 181–195, 2007.

[PRI 97] PRINCE R.C., "Bioremediation of marine oil spills", *Trends in Biotechnology*, vol. 15, no. 5, pp. 158–160, 1997.

[PRI 03] PRINCE, R.C., GARRETT R.M., BARE R.E. *et al.,* "The roles of photooxidation and biodegradation in long-term weathering of crude and heavy fuel oils", *Spill Science & Technology Bulletin*, vol. 8, no. 2, pp. 145–156, 2003.

[PRO 12] PROVENCHER J., GASTON A., O'HARA P. *et al.*, "Seabird diet indicates changing Arctic marine communities in eastern Canada", *Marine Ecology Progress Series*, vol. 454, pp. 171–182, doi: 10.3354/meps09299, 2012.

[RAM 04] RAMACHANDRAN S.D., HODSON P.V., KHAN C.W. *et al.*, "Oil dispersant increases PAH uptake by fish exposed to crude oil", *Ecotoxicology Environmental Safety*, vol. 59, pp. 300–308, 2004.

[RAM 10] RAMIREZ-LLODRA E., BRANDT A., DANOVARO R. *et al.*, "Deep, diverse and definitely different: unique attributes of the world's largest ecosystem", *Biogeosciences*, vol. 7, no. 9, pp. 2851–2899, 2010.

[RAM 11] RAMIREZ-LLODRA E., TYLER P.A., BAKER M.C. *et al.*, "Man and the last great wilderness: human impact on the deep sea", *PLoS One*, vol. 6, no. 8, p. e22588, 2011.

[ROJ 09] ROJO F., "Degradation of alkanes by bacteria", *Environmental Microbiology*, vol. 11, no. 10, pp. 2477–2490, 2009.

[RÖL 10] RÖLING W.F.M., "Hydrocarbon-degradation by acidophilic microorganisms", in TIMMIS K.N. (ed.), *Handbook of Hydrocarbon and Lipid Microbiology*, Springer, Berlin, Heidelberg, pp. 1923–1930, 2010.

[RON 02] RON E.Z., ROSENBERG E., "Biosurfactants and oil bioremediation", *Current Opinion in Biotechnology*, vol. 13, no. 3, pp. 249–252, 2002.

[SAN 00] SANTAS R., SANTAS P., "Effects of wave action on the bioremediation of crude oil saturated hydrocarbons", *Marine Pollution Bulletin*, vol. 40, no. 5, pp. 434–439, 2000.

[SCH 09] SCHLEICHER T., WERKMEISTER R., RUSS W. *et al.*, "Microbiological stability of biodiesel–diesel-mixtures", *Bioresource Technology*, vol. 100, no. 2, pp. 724–730, 2009.

[SCH 14] SCHEDLER M., HIESSL R., JUÁREZ A.G.V. *et al.*, "Effect of high pressure on hydrocarbon-degrading bacteria", *AMB Express*, vol. 4, no. 1, p. 77, 2014.

[SCH 17] SCHEDLER M., Microbial degradation of crude oil at high pressure, Doctoral dissertation, Technischen Universität Hamburg-Harburg, Rostock, 2017.

[SÉB 98] SÉBERT P., PERAGON J., BARROSO J.B. *et al.*, "High hydrostatic pressure (101 ATA) changes the metabolic design of yellow freshwater eel muscle", *Comparative Biochemistry and Physiology – Part B: Biochemistry and Molecular Biology*, vol. 121, no. 2, pp. 195–200, 1998.

[SHO 03] SHORT J.W., RICE S.D., HEINTZ R.A. *et al.*, "Long-term effects of crude oil on developing fish: lessons from the Exxon Valdez oil spill", *Energy Sources*, vol. 25, no. 6, pp. 509–517, 2003.

[SIM 06] SIMONATO F., CAMPANARO S., LAURO F.M. *et al.*, "Piezophilic adaptation: a genomic point of view", *Journal of Biotechnology*, vol. 126, no. 1, pp. 11–25, 2006.

[SME 17] SMEDILE F., LA CONO V., GENOVESE M. *et al.*, "High pressure cultivation of hydrocarbonoclastic aerobic bacteria", in MCGENITY T., TIMMIS K., NOGALES B. (eds), *Hydrocarbon and Lipid Microbiology Protocols, Springer Protocols Handbooks*, Springer, Berlin, pp. 173–173, doi.org/10.1007/8623_2017_231, 2017.

[SOM 91] SOMERO G.N., "Hydrostatic pressure and adaptations to the deep sea", in PROSSER C.L. (ed.), *Environmental and Metabolic Animal Physiology, Comparative Animal Physiology*, 4th ed., Wiley-LISS, New York, pp. 167–204, 1991.

[SOM 92] SOMERO G.N., "Adaptations to high hydrostatic pressure", *Annual Review of Physiology*, vol. 54, pp. 557–577, 1992.

[SOR 12] SOROKIN D.Y., JANSSEN A.J., MUYZER G., "Biodegradation potential of halo (alkali) philic prokaryotes", *Critical Reviews in Environmental Science and Technology*, vol. 42, no. 8, pp. 811–856, 2012.

[SPO 00] SPORMANN A.M., WIDDEL F., "Metabolism of alkylbenzenes, alkanes, and other hydrocarbons in anaerobic bacteria", *Biodegradation*, vol. 11, nos 2–3, pp. 85–105, 2000.

[STA 97] STANGE K., KLUNGSØYR J., "Organochlorine contaminants in fish and polycyclic aromatic in sediments from the Barents Sea", *ICES Journal of Marine Science*, vol. 54, no. 3, pp. 318–332, 1997.

[TAP 10] TAPILATU Y., ACQUAVIVA M., GUIGUE C. *et al.*, "Isolation of alkane-degrading bacteria from deep-sea Mediterranean sediments", *Letters in Applied Microbiology*, vol. 50, no. 2, pp. 234–236, 2010.

[TAR 16] TARRANT A.M., BAUMGARTNER M.F., LYSIAK N.S. *et al.*, "Transcriptional profiling of metabolic transitions during development and diapause preparation in the copepod Calanus finmarchicus", *Integrative and Comparative Biology*, vol. 56, no. 6, pp. 1157–1169, 2016.

[THE 00] THERON M., GUERRERO F., SÉBERT P., "Improvement in the efficiency of oxidative phosphorylation in the freshwater eel acclimated to 10.1 MPa hydrostatic pressure", *Journal of Experimental Biology*, vol. 203, no. 19, pp. 3019–3023, 2000.

[THE 14] THERON M., BADO-NILLES A., BEUVARD C. *et al.*, "Chemical fuel oil dispersion: acute effects on physiological, immune and anti-oxidant systems in juvenile turbot (Scophthalmus maximus)", *Water, Air, and Soil Pollution*, vol. 225, no. 3, p. 1887, doi: 10.1007/s11270-014-1887-z, 2014.

[THI 03] THISTLE D., "The deep-sea floor: an overview", in TYLER P.A. (ed.), *Ecosystems of the World, Ecosystems of the Deep Oceans*, Elsevier, Amsterdam, vol. 28, pp. 5–39, 2003.

[TIS 15] TISSIER F., DUSSAUZE M., LE FLOCH N. *et al.*, "Effect of dispersed crude oil on cardiac function in seabass Dicentrarchus labrax", *Chemosphere*, vol. 134, pp. 192–198, 2015.

[TOM 10] TOMANEK L., "Variation in the heat shock response and its implication for predicting the effect of global climate change on species' biogeographical distribution ranges and metabolic costs", *Journal of Experimental Biology*, vol. 213, pp. 971–979, 2010.

[TYA 11] TYAGI M., DA FONSECA M.M.R., DE CARVALHO C.C., "Bioaugmentation and biostimulation strategies to improve the effectiveness of bioremediation processes", *Biodegradation*, vol. 22, no. 2, pp. 231–241, 2011.

[USG 00] US GEOLOGICAL SURVEY (USGS), *World Petroleum Assessment 2000 – Description and Results*, US Geological Survey, Reston, 2000.

[VAN 03] VAN BEILEN J.B., LI Z., DUETZ W.A. *et al.*, "Diversity of alkane hydroxylase systems in the environment", *Oil and Gas Science and Technology*, vol. 58, no. 4, pp. 427–440, 2003.

[VEV 09] VEVERS W.F., DIXON D.R., DIXON L.R.J., "The role of hydrostatic pressure on developmental stages of Pomatoceros lamarcki (Polychaeta: Serpulidae) exposed to water accommodated fractions of crude oil and positive genotoxins at simulated depths of 1000–3000 m", *Environmental Pollution*, vol. 158, no. 5, pp. 1702–1709, 2009.

[WAR 97] WARTENA E.M.M., EVENSET A., Effects of the Komi oil spill 1994 in the Nenets Okrug, north-west Russia, Akvaplan-niva report APN514, 1997.

[WAS 11] WASSMANN P., DUARTE C.M., AGUSTI S. *et al.*, "Footprints of climate change in the Arctic marine ecosystem", *Global Change Biology*, vol. 17, no. 2, pp. 1235–1249, 2011.

[WHI 12] WHITE H.K., HSING P.Y., CHO W. *et al.*, "Impact of the Deepwater Horizon oil spill on a deep-water coral community in the Gulf of Mexico", *Proceedings of the National Academy of Sciences*, vol. 109, no. 50, pp. 20303–20308, 2012.

[WID 01] WIDDEL F., RABUS R., "Anaerobic biodegradation of saturated and aromatic hydrocarbons", *Current Opinion in Biotechnology*, vol. 12, no. 3, pp. 259–276, 2001.

[WIL 01] WILHELMS A., LARTER S.R., HEAD I. *et al.*, "Biodegradation of oil in uplifted basins prevented by deep-burial sterilization", *Nature*, vol. 411, no. 6841, p. 1034, 2001.

[WOL 98] WOLFE M.F., SCHLOSSER J.A., SCHWARTZ G.J.B. *et al.*, "Influence of dispersants on the bioavailability and trophic transfer of hydrocarbons in sediments from the Barents Sea", *ICES Journal of Marine Science*, vol. 54, no. 3, pp. 318–332, 1998.

[WRI 04] WRIGHT A.L., WEAVER R.W., "Fertilization and bioaugmentation for oil biodegradation in salt marsh mesocosms", *Water, Air, and Soil Pollution*, vol. 156, no. 1, pp. 229–240, 2004.

[ZEC 01] ZECCHINON L., CLAVERIE P., COLLINS T. *et al.*, "Did psychrophilic enzymes really win the challenge?", *Extremophiles*, vol. 5, no. 5, pp. 313–321, 2001.

6

Law Review on Accidental Marine Pollution

Ways of thinking and elements of knowledge about accidental marine pollution are discussed in this chapter. The discussion takes into account recent events, comparisons and prospective issues from the Prestige and Erika tanker accidents. Today, current affairs concern the Spanish Supreme Court on the Prestige tanker case, the Sanchi tanker accident in the China Sea and the 40th anniversary of the Amoco Cadiz sinking on the Brittany coast.

6.1. Introduction

Recently, on January 6th, 2018, a collision occurred between the Panamanian flag tanker Sanchi and a Chinese cargo ship in the China Sea. Once again, the marine environment's vulnerability is recalled to face the risk of marine pollution, particularly a hydrocarbon spill. A major pollution incident occurred with the leak of approximately 136,000 tons of light hydrocarbons. These condensates were burning, evaporating or leaking into the sea. This human tragedy raised several questions on maritime safety. The

---

Chapter written by Yann RABUTEAU and Frédéric MUTTIN.

tanker wreck was 120 m deep on the seabed, to which we have to add the remaining heavy oil for the ship's propulsion.

Besides this new disaster, let us remember the last major oil spills that affected the European shoreline and the recent development concerning the rights on the pollution of the seas. At the same time, in France, we will soon commemorate the 40th anniversary of the catastrophe and the pollution of the tanker Amoco Cadiz [CED 18]. The tragedy took place on March 16th, 1978 off the Portsall coast in the Finistere county.

On the night of November 13th, 2002, the Bahamian flaged tanker Prestige was out of control off the Galician coasts, and finally the hull of the ship broke into two parts after several towing attempts. As a result, there was a heavy oil leakage (64,000 tons out of the oil freight 77,000 tons), and the tanker wreck sank to a depth of 3,000 m.

Looking at the French and Spanish Atlantic coasts, that event is the second largest oil spill after the wreckage and the pollution of the Erika tanker. In both cases, the ship was a 15-year-old single-hull tanker that carried heavy oil and broke into two parts after structural damage. As a result, these events caused major pollution and large and intense impacts on the Spanish shoreline around the city of A Coruña. These disasters also affected the French coast with lower intensity, from the Bay of Biscay to the Finistere county, as well as north-west Portugal.

The pollution resulted in very large operations at sea, the aim of which was to recover most of the hydrocarbons before their successive oil beaching. On the other hand, several attempts were made to recover and contain the remaining oil in the two parts of the hull. These offshore operations involved very high costs. The costs of the cleaning operations must therefore be added. The pollution particularly affected the Galician region. The damage caused was considerably greater, thereby affecting the coastal and marine environments as well as devastating the flora and fauna with significant losses of seabirds as well as specific shellfish and crustacean deposits. More precisely, in France, Spain and Portugal, the pollution affected 2,900 km of their coastline. According to a report, approximately 115,000 to 230,000 seabirds died [CED 14].

Economic harm must also be considered as a consequence of the oil spill, which mainly concerned the tourism and fishing industries. The exploitation of sea products was already considered, including sea angling. In the Galicia region, the traditional collection of barnacles (*Pollicipes pollicipes*) had been practiced on the shore. However, the Galician authority prohibited recreational or subsistence fishing for a long time as a result of shellfish contamination caused by hydrocarbon pollution on the shore.

6.2. Injury and compensation

On the one hand, the total cost amounting to 4,000,000,000 € (4 billion euros) represented the injuries caused by the pollution of the Prestige[1]; however, on the other hand, the total fund available was 171,500,000 € (171.5 million euros) under the civil liability of the ship owner in the framework of the 1992 CLC Convention[2] [CLC 92] and

1 In the framework of the different legal procedures ongoing in France and Spain, this amount corresponds to the whole set of injuries delivered to the ship's insurance company.

2 The Civil Liability Convention (1992 CLC) governs the liability of ship owners for oil pollution damage. Under this convention, the registered ship owner has strict liability for pollution damage caused by the escape or discharge of persistent oil from his ship. The ship owner is normally entitled to limit his liability to an amount determined by the size of the ship (from: http://www.iopcfunds.org/about-us/legal-framework/1992-civil-liability-convention).

the IOPC[3] Fund Convention of 1992 [IOP 92]. The latter amount included the compensation given by the insurance (P&I club).

The limitation on the civil liability applicable to the ship owner came from international conventions. For this reason, the victims decided to sue the polluters because they failed to reach a friendly agreement. Two objectives were put forward. The first objective was to demonstrate the offense in the occurrence of the damage and shipwreck. This permits the assignment of civil liability of other parties and not just the ship owner. The second objective was to admit the reality and the admissibility of injuries other than those acceptable in the framework of the 1992 CLC Convention and the 1992 IOPC Fund. Notably, the compensation demands address the other issues as follows: non-material moral damage, environmental losses and ecological loss. The two injured states were also engaged in penal claims. These concerned the occurrence of an accidental marine pollution that had a negative impact on the environment and a protected natural area. The head of the Spanish Maritime Transportation Administration, the Greek captain of the ship, the second officer and the engineering officer were directly sued during the judiciary action.

The judicial process and investigation were exclusively carried out by the Spanish law administration. Meanwhile, some judicial actions were initiated in France at a preliminary level. In October 2012, ten years after the fact, a trial took place at the Court of A Coruña.

3 The 1992 Fund Convention (IOPC Fund), which is supplementary to the 1992 CLC, establishes a regime for compensating victims when compensation under the 1992 CLC is not available or is inadequate. The International Oil Pollution Compensation Fund, 1992 (1992 Fund), was set up under the 1992 Fund Convention (from: http://www.iopcfunds.org /about-us/legal-framework/1992-fund-convention-and-supplementary-fund-protoco).

On November 13th, 2013, a first lawsuit was taken, which surprised the pollution victims. Indeed, all of the accused parties were acquitted on all charges by the court, except the ship's captain who was sentenced because he refused to have his ship towed. A new judicial process took place at the Spanish Supreme Court following this first sentence. On January 26th, 2016, the Supreme Court gave its decision. Finally, the ship's captain, the ship's owner and equally the civil liability insurance company of the ship were found guilty.

Nevertheless, a new judicial session was necessary at the Audiencia Provincial A Coruña. It allowed the final amount of all financial compensations to be fixed. It then had to be considered by the Supreme Court[4]:

> *"to assess the repairing effort of the finding damages, the loss of earning, including the damages to the environment in all its different forms, the compensation of the physical and moral damages, within the limits of the request formulated by the parties in their conclusions".*

6.3. Court decision and case law

The final decision on the Prestige case was made on November 15th, 2017. The total compensation estimation was 1,573,000,000 € (1.573 billion euros) for the oil spill damages. The total cost required by the victims' parties was 4,000,000,000 € (4 billion euros). France would receive 61,000,000 € (61 million euros). This compensation would be shared between the state, the local authorities and the private victims. The insurance company of the tanker and the IOPC Fund had to pay complementary indemnities. Nevertheless, the Spanish jurisdiction and the international conventions dedicated to the compensation of marine pollution avoided the real ecological cost. No indemnity

4 Warrant pronounced on January 26th, 2016.

decision was made to compensate for the mortality of fauna and the damage to flora.

Later on, this decision was assimilated to the one given by the Supreme Court of France (*Cour de Cassation*) in 2012. This concerned the final decision on the Erika affair, which sank on December 12th, 1999 near Penmarc'h cape, 70 km south, Finistere, France.

The facts between the two cases appear similar (notably, the same kind of ship, same oil freight, similar weather conditions and lack of good care of the ship). From the judicial perspective, several similarities exist in the duration and the court process. However, the judicial issues between the two cases are seriously different.

The Supreme Court of France has recognized for "the first time" the need to compensate for the ecological loss in a maritime affair. This prejudice is unrelated and different from other damages resulting from an oil spill. It must also be financially counterbalanced. However, the judges in both states pronounced penal and civil offenses against the ship's owners, the tanker's freighters and the classification company.

This specific ecological loss amounted to 13,000,000 € (13 million euros). This compensation was part of the total fine decided by the court for the damage costs 200,600,000 € (200.6 million euros). After this recognition by the case law, the ecological loss concept was acknowledged by the legislator. It was formally incorporated into the French civil law in 2016[5].

5 Code Civil, France
Art. 1386-19.-Toute personne responsable d'un préjudice écologique est tenue de le réparer [COD 16a], i.e. "Anyone accountable for an ecological loss owes compensation".
Art. 1386-20.-Est réparable, dans les conditions prévues au présent titre, le préjudice écologique consistant en une atteinte non négligeable aux éléments ou aux fonctions des écosystèmes ou aux bénéfices collectifs tirés par l'homme de l'environnement [COD 16b], i.e. "is compensable, in the

6.4. Discussion

We will now move our discussion beyond this comparison of the two decisions. To this end, let us define the remaining difficulty with the compensation of the damages to the marine environment and the ecological loss due to marine pollution.

The legal recognition of maritime ecological loss has recently been affirmed in the Erika affair. This is important; however, restoring ecological damage in practice remains a complex issue. The lack of specific data and economical assessment methods of the ecological loss may explain the statement in the Prestige affair. The specificity of the judicial elements in the Spanish Code may stay outside of this interpretation. Let us consider the case where a legal assessment exists for environmental damage. The compensation can equally be made in kind or with non-financial means.

Nevertheless, the proof of ecological damage, and especially the assessment of this damage, remains today an obstacle to a fair compensation. In both cases, it is fundamental to obtain data on the environmental initial state and the monitoring of affected zones for their qualitative properties. This issue is notably for coastal waters, harbor waters and the biodiversity in order to dispose of appraising elements, which have to be as rigorous as possible. These elements must be evaluated during the damage assessment and consequently in line with restoring actions.

For this purpose, the initiated research projects [BRE 16] can be very helpful for the judges. These projects can monitor, for example, a natural species taken as a reference for the environment. It allows the judicial procedure to distinguish the effects between accidental pollution and

condition of this article, the ecological loss as being a significant injury to the elements and other functions of the ecosystems or to the nature's services granted to people".

chronic pollution. Moreover, it is valuable for choosing the compensation means when damages occur in the natural marine and coastal environments.

6.5. Conclusion

In the future, it is necessary to work on the definition of assessment methods dedicated to injuries, damage appraisal and compensation regimes. It must integrate International Law or European Union Law. In addition, it should be made necessary to activate a definition of Universal Ecological Loss in the case of marine pollution. This issue is not always claimed.

6.6. Bibliography

[BRE 16] BREITWIESER M., VIRICEL A., GRABER M. *et al.*, "Reaching for integrative bio-monitoring tools to assess chronic chemical contamination in the coastal environment", *ICES MSEAS 2016*, Brest, France, 2 June 2016.

[CED 14] CEDRE, Prestige, http://wwz.cedre.fr/Ressources/Accidentologie/Accidents/Prestige, 1 April 2014.

[CED 18] CEDRE, "Amoco Cadiz, 40 ans d'évolution(s) (2018)", *Journée d'information du Cedre*, Brest, France, Conclusions de la journée du 16 mars 2018, pp. 1–3, available at: https://wwz.cedre.fr/content/download/9268/147396/file/conlusions-40-ans-amoco.pdf, 16 March 2018.

[CLC 92] CLC-CONVENTION, http://www.iopcfunds.org/about-us/legal-framework/1992-civil-liability-convention, 1992.

[COD 16a] CODE CIVIL – Article 1386-19. Created by the LOI n°2016-1087 of 8 August 2016, Legifrance, https://www.legifrance.gouv.fr, 2016.

[COD 16b] CODE CIVIL – Article 1386-20. Created by the LOI n°2016-1087 of 8 August 2016, Legifrance, https://www.legifrance.gouv.fr, 2016.

[IOP 92] IOPC-FUND, http://www.iopcfunds.org/about-us/legal-framework/1992-fund-convention-and-supplementary-fund-protoco, 1992.

List of Authors

Marine BREITWIESER
LIENSs CNRS
La Rochelle University
France

Rose CAMPBELL
EIGSI
La Rochelle
France

Matthieu DUSSAUZE
LMGE CNRS
Clermont-Auvergne University
Clermont-Ferrand
France

Angélique FONTANAUD
Marina Harbor
La Rochelle
France

Stéphane LE FLOCH
CEDRE
Brest
France

Florian LELCHAT
LABOCEA
Plouzané
France

Thomas MILINKOVITCH
LIENSs CNRS
La Rochelle University
France

Frédéric MUTTIN
EIGSI
La Rochelle
France

Yann RABUTEAU
ALLEGANS
University of Western Brittany
Brest
France

Hélène THOMAS-GUYON
LIENSs CNRS
La Rochelle University
France

Index

C

D, E, F

G

R, S, T

V, W, X, Z

Printed in the United States
By Bookmasters